Plasmonic Nanomaterials

Plasmonic Nanomaterials

Characterization and Applications in Organic Synthesis and Catalysis

edited by

Clémence Queffélec
D. Andrew Knight

JENNY STANFORD
PUBLISHING

Published by

Jenny Stanford Publishing Pte. Ltd.
101 Thomson Road
#06-01, United Square
Singapore 307591

Email: editorial@jennystanford.com
Web: www.jennystanford.com

British Library Cataloguing-in-Publication Data
A catalogue record for this book is available from the British Library.

Plasmonic Nanomaterials: Characterization and Applications in Organic Synthesis and Catalysis

Cover image courtesy: VR2Planets

ISBN 978-981-5129-09-0 (Hardcover)
ISBN 978-1-003-47406-7 (eBook)

Contents

Preface

In recent years, the field of nanoscience has witnessed remarkable progress, leading to unprecedented advancements in various scientific disciplines. Among these, plasmonic nanomaterials have emerged as a cornerstone in the fields of materials science and catalysis. Their unique optical properties and surface-enhanced reactivity have revolutionized our approach to organic synthesis and catalytic processes.

This book stands at the forefront of this exciting and rapidly evolving field. It is designed to serve as a comprehensive guide for researchers, students, and professionals seeking a deeper understanding of the principles underlying the characterization and utilization of plasmonic nanomaterials in the context of organic synthesis and catalysis.

The book provides a thorough exploration of the fundamental principles governing the behavior of plasmonic nanomaterials, coupled with practical insights into their characterization techniques and applications in catalytic processes. Through a meticulous blend of theoretical concepts and experimental methodologies, this book strives to equip readers with the knowledge and skills necessary to harness the full potential of these extraordinary materials.

From the basics of plasmonics to advanced characterization techniques, from the design principles of plasmonic catalysts to their applications in diverse organic transformations, this book endeavors to be a valuable resource for both novices entering the field as well as researchers looking to expand their expertise.

We are indebted to the dedicated scientists and researchers whose pioneering work in the field of plasmonic nanomaterials has paved the way for the content presented in this book. Their groundbreaking contributions have not only enriched our understanding but have also opened up new frontiers in the domain of catalysis.

We hope this book will initiate a wide-ranging discussion about the potential and challenges associated with the use of plasmonic nanomaterials for catalysis. It is our sincere belief that the knowledge contained within these pages will inspire and empower the next generation of scientists and engineers to push the boundaries of what is possible in the field of organic synthesis and catalysis.

Clémence Queffélec
D. Andrew Knight
Autumn 2023

Chapter 1

Introduction to Plasmonics

Wafa Safar, Mathieu Edely, and
Marc Lamy de la Chapelle
Le Mans University, Laboratoire IMMM, UMR6283,
Avenue Olivier Messiaen 72085 Le Mans Cedex 9, France
mathieu.edely@univ-lemans.fr

A plasmon is an optical characteristic of a metal resulting from the excitation of free electrons by their interaction with photons or electrons [1, 2]. As these free electrons are weakly bounded to the metal atom, the plasmon corresponds to a collective oscillation of the charge density, named "plasma oscillation." It occurs when photons or electrons transfer their energy to the free electrons of the metal since their momentum is the same as that of the plasmon. Depending on the geometrical dimensions of the metallic materials, plasmons can be defined and classified as follows:

- **Volume plasmon (3D plasmon):** The plasmon is excited in the material volume.
- **Delocalized surface plasmon (DSP), also known as surface plasmon polariton (SPP) (2D plasmon):** The plasmon is formed at the interface between a dielectric and a metal.

Plasmonic Nanomaterials: Characterization and Applications in Organic Synthesis and Catalysis
Edited by Clémence Queffélec and D. Andrew Knight

ISBN 978-981-5129-09-0 (Hardcover), 978-1-003-47406-7 (eBook)
www.jennystanford.com

Having an evanescent character, it can only propagate along the surface.

- **Localized surface plasmon (LSP) (0D plasmon):** The plasmon is confined inside a nanostructure. It can be seen as an eigen mode of oscillation of the free electrons inside a limited volume. Different plasmon modes can be excited depending on the size, shape, and chemical nature of the nanostructure.

1.1 Dielectric Constant and Optical Properties of Materials

To understand plasmons and their features, it is valuable to describe the relation between the optical and electronic properties of a metal. The interaction between the light and the plasmon corresponds to the optical part, whereas the electronic properties refer to the description of the behavior of the free electrons located in the conduction bands of the metal. Several models have been proposed to simulate and explain the electronic properties and their relations with the material. They are based on the definition of the interactions of the electrons with other metal elements, such as the ion array, as well as on the nature of the electrons. Three different models have been proposed:

1. **Drude model (1900):** The electrons interact only with the metallic ions through collisions [3, 4].
2. **Drude–Lorentz model (1905):** Seen as an extension of the Drude model, the electrons are also subject to a restoring force through its electrostatic interaction with the metallic ions [5, 6].
3. **Drude–Sommerfeld model:** As the electron is a fermion and a quantic object, it is submitted to the Fermi–Dirac distribution, which has a consequence on its behavior.

In the following, we will present only the Drude–Lorentz model, which is enough to describe the plasmon behavior in the electromagnetic formalism and the interaction between an electromagnetic wave and a material [5, 6].

The material is then described by its optical properties, particularly its dielectric constant ε. It depends on the excitation frequency and determines the metal's response to an electromagnetic radiation:

$$\varepsilon = \varepsilon' + i\varepsilon'' \quad (1.1)$$

The real part of ε represents the effect of the electrical field on the polarization of the material. The imaginary part represents the damping of the radiation and the losses within the material.

In 1900, Drude first proposed to describe electrons as a gas (the Drude model). The free electrons do not interact with each other but can interact with the atom's nucleus considered as ions. This interaction between the ions and the electrons is defined as the probability of collision, which induces the damping of the electron motion and is modeled thanks to the viscosity force F_D defined by:

$$\boldsymbol{F}_{\mathbf{D}} = -\frac{m}{\tau}\nu \quad (1.2)$$

where τ is the average time between two collisions (defined as relaxation time), ν is the electron velocity, and m is the electron mass.

Lorentz later refined this model (the Drude–Lorentz model) in 1905 by introducing the electrostatic interaction of the electrons with the positively charged ions corresponding to the oscillation of the electrons around an equilibrium position. The electron is then subject to the elastic force F_L described by:

$$F_L = -Kx \quad (1.3)$$

where K is the elastic constant and x is the electron position.

As the electrons have an electric charge $-e$, with $e = 1.602 \times 10^{-19}$ C, they are also subject to an oscillating electric field when they are excited by an electromagnetic wave and thus to a force F_C:

$$F_C = -eE = -eE_0e^{-i\omega t} \quad (1.4)$$

where $E = E_0(\omega)e^{-i\omega t}$ is the excitation electric field, E_0 is its amplitude, and ω is the frequency of the oscillating electric field.

The electron motion can be calculated using the fundamental principle of dynamics, as follows:

$$m\ddot{x} = F_D + F_L + F_C = -\frac{m}{\tau}\dot{x} - Kx - eE_0e^{-i\omega t} \quad (1.5)$$

or $$\ddot{x}+\frac{1}{\tau}\dot{x}+\frac{K}{m}x=-eE_0e^{-i\omega t}$$

This differential equation can be solved easily, and the electron motion can be described as:

$$x(t) = x_0(\omega)\, e^{-i\omega t} \text{ and } x_0(\omega)=\frac{eE_0}{m\left(\omega^2-\omega_0^2+i\frac{\omega}{\tau}\right)} \tag{1.6}$$

where $\omega_0^2 = \frac{K}{m}$, the eigen frequency.

Considering this solution, the electrons will be submitted to an oscillation. The frequency of oscillation ω will be the same as that of the excitation electric field. Moreover, the oscillation amplitude x_0 depends on the excitation frequency and the material characteristics (K, m, τ). This oscillation amplitude is maximum at the ω_0 pulsation, which means that the absorption of light will be maximum and will reach a resonance condition.

The electron motion induces a dipolar moment $\vec{P}$ in the material, proportional to the electron motion, $x(t)$, and given by the direction of the electrical field $\vec{E}$:

$$\vec{P} = -n_e\, e\, x(t)\,\vec{u} = -n_e\, e\frac{x_0}{E_0}\vec{E} \tag{1.7}$$

where e is the electron charge, n_e is the electron density inside the material, and $\vec{u}$ is the unitary vector in the motion direction.

$\vec{P}$ can also be written as:

$$\vec{P} = \varepsilon_0\chi_e\vec{E} = \varepsilon_0 N\alpha\vec{E} \tag{1.8}$$

where ε_0 is the dielectric constant in the vacuum (ε_0= 8.85418782 × $10^{-12}A^2s^4\,kg^{-1}m^{-3}$), χ_e is the electrical susceptibility, N is the number of dipoles, and α is the polarizability.

The electric induction in the material is known as:

$$\vec{D}=\varepsilon_0\vec{E}+\vec{P} = \varepsilon_0\vec{E}+\varepsilon_0 N\alpha\vec{E} \tag{1.9}$$

$$\vec{D} = \varepsilon_0\varepsilon_r\vec{E} = \varepsilon_0(1+\chi_e)\vec{E} \tag{1.10}$$

where ε_r is the relative dielectric constant of the material.

One can then deduce the following from Eqs. (1.6), (1.7), (1.9), and (1.10):

$$\varepsilon_r = 1 + \chi_e = 1 + n_e\, e\alpha = 1 - \frac{n_e e x_0(\omega)}{\varepsilon_0 E_0} \tag{1.11}$$

$$\varepsilon_r = 1 - \frac{\omega_p^2}{\omega^2 - \omega_0^2 + i\dfrac{\omega}{\tau}} \tag{1.12}$$

where $\omega_p^2 = \dfrac{n_e e^2}{m\varepsilon_0}$ is the plasma frequency.

In the ideal case, the free electron gas is without damping and has an ω_0 equal to 0 as described by the Drude model. The relative dielectric constant can be approximated by:

$$\varepsilon_r = 1 - \frac{\omega_p^2}{\omega^2 + i\dfrac{\omega}{\tau}} = 1 - \frac{\omega_p^2}{\omega^2} \tag{1.13}$$

When $\varepsilon_r > 0$, the electrons are displaced in the opposite direction of E. The field polarizes the material that acts as a dielectric.

When $\varepsilon_r < 0$, the electrons are displaced in the direction of E and create a depolarized field in the material that acts as a metal.

1.2 Volume Plasmons

In 1920, Rudberg excited metals with an electron beam and measured the energies of the electrons after their interaction with the metal [7]. He noticed losses in electron energy at specific energy levels. The discretization of the energy lost by the free electrons of the metal reflects the quantification of the plasma oscillation energy. The quantum of the collective excitation of electron gases was then called plasmon, and its energy is a multiple of $\hbar\omega_p$. In the bulk metal, the electrons then oscillate, collectively forming a plasmon. This plasmon is infinitely expanded in the metal as the entire electron cloud oscillates at the plasma frequency ω_p in the metal volume.

1.3 Surface Plasmons

1.3.1 Plasmon Dispersion Curve

Surface plasmon can also be excited at the interface between a dielectric medium and a metal, and an electromagnetic wave is formed, which can propagate at the surface of the metal. Such a plasmon is known as delocalized surface plasmon (DSP) or surface plasmon polariton (SPP) [1, 2, 8]. The SPP can be excited using a transverse magnetic (TM) polarized electromagnetic wave. The k-vector of the excitation light has also to be equal the SPP k-vector to reach the phase matching between both k-vectors.

Let us consider an electromagnetic wave at a dielectric/metal interface as presented in Fig. 1.1 with ε_m and ε_d, the dielectric constants of the metal and the dielectric medium, respectively.

The two electric fields in the dielectric and metallic media (E_d and E_m, respectively) can be written as:

$$E_d = (E_{xd}, 0, E_{zd})\, e^{i(k_{xd}x + k_{zd}z - \omega t)} \tag{1.14}$$

$$E_m = (E_{xm}, 0, E_{zm})\, e^{i(k_{xm}x + k_{zm}z - \omega t)} \tag{1.15}$$

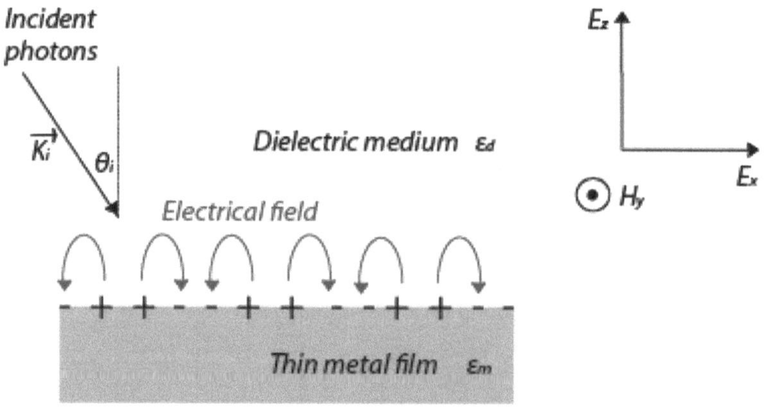

Figure 1.1 A schematic representing the excitation of a delocalized surface plasmon at the interface between a metal and a dielectric. ε_m and ε_d are the dielectric constants of the metal and the dielectric medium, respectively.

Using Maxwell's equations and the field continuity relations at the interface, the fields can be expressed as:

$$E_{xd} = E_{xm} \text{ and } H_{xd} = H_{xm} \tag{1.16}$$

$$D_{zd} = \varepsilon_d E_{zd} = \varepsilon_m E_{zm} = D_{zm} \tag{1.17}$$

$$\frac{k_{\mathrm{zd}}}{\varepsilon_{\mathrm{d}}}+\frac{k_{\mathrm{zm}}}{\varepsilon_{\mathrm{m}}}=0 \tag{1.18}$$

One can notice a discontinuity of the component of the electric field along the z-direction (perpendicular to the surface). This discontinuity will be at the origin of the plasmon excitation, and this explains why we need to have a TM-polarized electromagnetic wave to create the plasmon.

The dispersion relation of light can be expressed in the dielectric or metal media as:

$$k_{\mathrm{xm}}^2+k_{\mathrm{zm}}^2=\varepsilon_{\mathrm{m}}\left(\frac{\omega}{c}\right)^2 \tag{1.19}$$

$$k_{\mathrm{xd}}^2+k_{\mathrm{zd}}^2=\varepsilon_{\mathrm{d}}\left(\frac{\omega}{c}\right)^2 \tag{1.20}$$

Assuming that the k-vectors are equal in the x direction ($k_{\mathrm{xd}} = k_{\mathrm{xm}} = k_{\mathrm{x}} = k_{\mathrm{SP}}$), and that, from Eq. (1.18):

$$k_{\mathrm{zm}}^2=\left(\frac{\varepsilon_{\mathrm{m}}}{\varepsilon_{\mathrm{d}}}\right)^2 k_{\mathrm{zd}}^2 \tag{1.21}$$

one can deduce that:

$$k_{\mathrm{zm}}^2=\varepsilon_{\mathrm{m}}\left(\frac{\omega}{c}\right)^2-k_x^2=\left(\frac{\varepsilon_{\mathrm{m}}}{\varepsilon_{\mathrm{d}}}\right)^2 k_{\mathrm{zd}}^2=\left(\frac{\varepsilon_{\mathrm{m}}}{\varepsilon_{\mathrm{d}}}\right)^2\left[\varepsilon_{\mathrm{d}}\left(\frac{\omega}{c}\right)^2-k_x^2\right] \tag{1.22}$$

and that the dispersion relation of the SPP is:

$$\omega=c\,k_{\mathrm{x}}\sqrt{\frac{\varepsilon_{\mathrm{m}}+\varepsilon_{\mathrm{d}}}{\varepsilon_{\mathrm{m}}\varepsilon_{\mathrm{d}}}} \tag{1.23}$$

where c is the light velocity in the vacuum and ω is the SPP frequency. Referring to the Drude model and replacing ε_{m} with the ε_{r} calculated previously (Eq. 1.13):

$$\omega=\frac{1}{\sqrt{2\varepsilon_{\mathrm{d}}}}\left(\varepsilon_{\mathrm{d}}\omega_p^2+(1+\varepsilon_{\mathrm{d}})c^2k_x^2-\left(\left(\varepsilon_{\mathrm{d}}\omega_p^2+(1+\varepsilon_{\mathrm{d}})c^2k_x^2\right)^2-4\varepsilon_{\mathrm{d}}\omega_p^2c^2k_x^2\right)^{\frac{1}{2}}\right)^{\frac{1}{2}} \tag{1.24}$$

For low values of the wave vector k_x, the plasmon dispersion curve increases nearly linearly, whereas for high values of k_x, the curve reaches a plateau and becomes tangential to the plasma frequency $\omega_p/\sqrt{1+\varepsilon_d}$.

1.3.2 Surface Plasmon Characteristics

As the dielectric constant of metal is negative ($\varepsilon_m < 0$), the plasmon frequency is always lower than the light frequency in the dielectric medium or in the metal:

$$\omega_{\text{Plasmon}} = ck_x\sqrt{\frac{\varepsilon_m+\varepsilon_d}{\varepsilon_m\varepsilon_d}} < \frac{1}{\sqrt{\varepsilon_{\text{m or d}}}}ck_x = \omega_{\text{Light}} \tag{1.25}$$

The dispersion curve of the surface plasmon is then always below the dispersion curve of the light (Fig. 1.2).

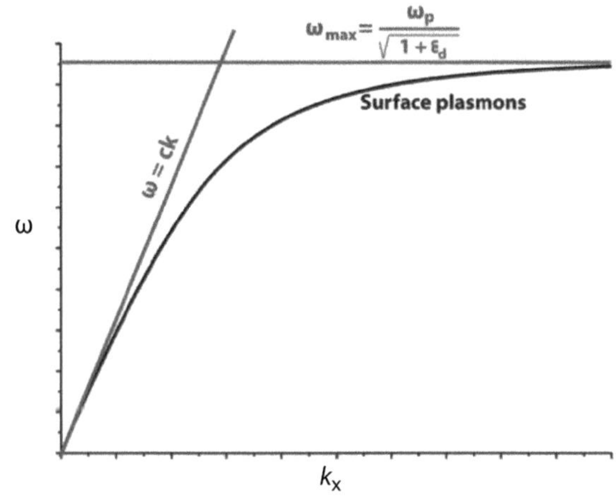

Figure 1.2 Dispersion curve of the surface plasmon polariton at the interface between the metal and the dielectric medium.

If we consider the wave vector in the z-direction for both materials (dielectric medium and metal), we find from Eqs. (1.19), (1.20), and (1.25) that:

$$k_{\text{zm or zd}}^2 = \varepsilon_{\text{m or d}}\left(\frac{\omega}{c}\right)^2 - k_x^2 < 0 \tag{1.26}$$

The wave vector in the z-direction is then imaginary in both dielectric and metal media. This means that the surface plasmon is an evanescent wave and that its amplitude decreases exponentially in the direction perpendicular to the metallic surface. The plasmon is then confined at the dielectric–metal interface. The decay length of the plasmon perpendicular to the surface is between 10 and 50 nm in the metal and about 100–200 nm in the dielectric media.

However, the SPP can propagate along the surface at a distance of several tens of micrometers depending on the damping in the metal. The propagative length, L_{SP}, is determined by the imaginary part of the wave vector in the surface direction such as:

$$L_{SP} = \frac{1}{2Im(k_x)} \tag{1.27}$$

For instance, the values of L_{SP} are round 22 μm and 500 μm for silver at 514 and 1060 nm, respectively. The plasmon is then delocalized in the plane of the interface between the dielectric medium and the metal but is confined (evanescent wave) in the direction perpendicular to the interface.

1.3.3 Surface Plasmon Excitation

To excite delocalized surface plasmons, both frequencies and wave vectors of the light and the plasmon have to be equal (phase matching) and thus the light dispersion curve $\omega = ck$ has to intercept the SPP dispersion curve. However, the light dispersion curve is always above the SPP one (Fig. 1.2). To reach the curve intersection, the light wave vector value has to be increased.

1.3.3.1 Plasmon excitation using a grating

One way to increase the wave vector is to include a grating with a grating constant of p on the metal surface. The grazing orders of diffraction of the grating will be added to the excitation wave vector $k_{x\text{light}}$. The wave vector at the interface becomes:

$$k_x = k_{\text{light}} \sin\theta \pm m\frac{2\pi}{p} \tag{1.26}$$

where θ is the incidence angle of the excitation light and m is the order of diffraction. The value of k_x can then be increased by the diffraction order wave vector. It will induce a shift in the light curve, which can then intercept the plasmon dispersion curve to excite the plasmon [9].

1.3.3.2 Plasmon excitation using a prism

Another way is to increase the refractive index of the dielectric medium at the interface with the metal. Otto [10] and Kretschmann [11] proposed two different configurations to realize the light/SPP phase matching by using a prism close to the metal surface.

The Otto configuration was the first one to be proposed to excite the SPP. It uses a metallic film placed close but not in contact with the prism interface (Fig. 1.3a) forming an air layer between the prism and the metal surfaces. The excitation light goes through the prism with an angle to be in total internal reflection. An evanescent wave is created in the air at the prism surface. This evanescent wave will then be used to excite the SPP at the metal–air interface (Fig. 1.3a). This configuration is difficult to realize as it is hard to control a thin and homogeneous distance between the prism and the metal.

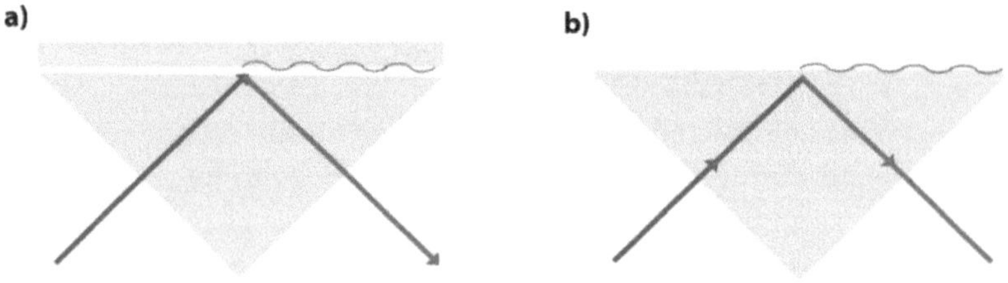

Figure 1.3 Schemes of (a) the Otto configuration and (b) the Kretschmann configuration.

For the Kretschmann configuration, a thin metallic film is deposited directly on the prism. As for the Otto configuration, the excitation light goes through the prism with an angle of total reflection at the prism–metal interface. An evanescent wave is created through the metal film that excites the metal in the opposite side of the film at the air–metal interface (Fig. 1.3b). The Kretschmann configuration is easier to realize than the Otto configuration, and thus it is the one currently used to excite SPP.

With such Kretschmann configuration, the surface plasmon can be determined easily. A laser beam is used to illuminate the metal film with a fixed excitation wavelength, and the reflected intensity is measured. The wave vector at the interface between the metal and the prism is defined as:

$$k_x = n_{prism}\, k_{light} \sin\theta \qquad (1.27)$$

where θ is the incidence angle of the excitation light, k_{light} is the incident light wavevector, and n_{prism} is the optical index of the prism. As k_{light} is fixed by the laser wavelength and n_{prism} is constant, k_x can be modified by changing the incident angle θ. By varying θ, one can reach the phase matching between the excitation light and the plasmon. In this case, one can observe a drastic decrease in the reflected intensity as the light energy is absorbed by the metal and transferred to the plasmon process. This intensity decrease is known as surface plasmon resonance (SPR). For instance, a gold thin film with a thickness of 50 nm and deposited on a glass prism (n = 1.5) exhibits an SPR for an incident angle close to 70° with an excitation wavelength of 660 nm.

From Eq. (1.23), one can notice that the plasmon dispersion curve depends on the dielectric constant of the dielectric medium above the gold film. Thus, if we modify this dielectric constant through the absorption or the grafting of molecules at the surface of the film, we will observe a shift in the dispersion curve and as a consequence of the SPR. This concept is at the basis of the SPR-sensing method [12, 13]. This method is essentially used to measure the affinity constant between two molecules: One is grafted at the gold surface, whereas the other is injected at different concentrations to characterize the association or dissociation constants of the molecular interaction. The local modification of the dielectric constant can be quantified by measuring the angular shift or the variation in the reflectivity intensity and correlated with the number of molecules deposited at the gold surface. This method can be qualitative (interaction or not) and quantitative (number of interacting molecules) and is largely used in biology to observe antibody/antigen interactions.

1.4 Localized Surface Plasmon

When a nanostructure such as a nanoparticle is excited by an electromagnetic wave, the penetration depth of the electromagnetic field is equal to or higher than the nanoparticle size. The plasmon can no longer propagate over large distances, as in the case of the delocalized surface plasmon. It is then confined inside the nanostructure volume and becomes a localized oscillation. In this case, it is known as localized surface plasmon (LSP) and can be observed on small structures that can trap the surface plasmons, such as nanocylinders, nanospheres, nanorods, nanoholes, or even topographical singularities on surfaces [14]. Thus, the LSP corresponds to an eigen mode of oscillation of the free electrons inside the nanostructures and is confined in their three dimensions. When the light frequency is equal to the eigen mode frequency, a resonance effect occurs and the LSP is excited. The position of the LSP resonance (LSPR) depends not only on the nature of the metal and the direct environment of the nanostructure (i.e., the surrounding medium), but also on the size (Fig. 1.4), shape, and spatial distribution of metallic nanostructures (coupled with other nanostructures). For instance, Fig. 1.4 shows that the plasmon resonance of nanocylinders is redshifted when the nanocylinder diameter increases from 100 to 200 nm [15].

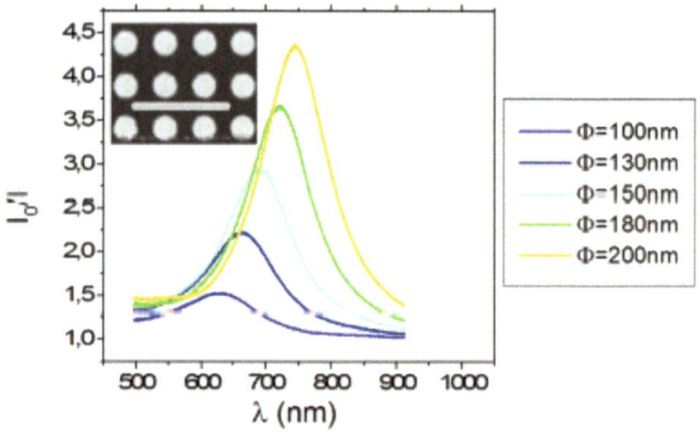

Figure 1.4 Extinction spectra of gold nanocylinders on glass surface. Diameters varying between 100 nm and 200 nm, height equals to 50 nm, and the gap between two nanocylinders is 200 nm in both directions (extracted from Ref. [15]).

The LSPR can be measured by extinction spectroscopy. The extinction spectrum corresponds to the intensity attenuation experienced by a light beam after interaction with a nanostructure. The extinction process results in two kinds of light interaction with the nanostructures: the absorption and the scattering of the incident light by the material. These two processes are also known as nonradiative and radiative interactions, respectively, as photons may or may not be emitted during the process.

The extinction cross section σ_{ext} is then defined as:

$$\sigma_{ext} = \sigma_{abs} + \sigma_{scat} \quad (1.27)$$

where σ_{abs} and σ_{scat} are the absorption and scattering cross sections of the material embedded in a dielectric medium. When plasmon is excited, the extinction spectrum exhibits a large band (width of a few tens of nanometers). The maximum of this band corresponds to the largest interaction of light with the nanoparticles and thus to the plasmon resonance. In the following, we will focus on the plasmon resonance of different kinds of nanostructures (nanospheres, elongated nanoparticles) and also on the influence of the different geometrical (size and shape), chemical (nature of the metal), and environmental (surrounding medium, coupling) parameters on the position of the plasmon resonance.

1.4.1 LSP in Spherical Nanoparticles

The electromagnetic wave interaction with the spherical particles can be completely determined using the Mie theory [16–20]. It solves analytically Maxwell's equation in the spherical coordinate system to calculate the extinction and scattering cross sections (the absorption one is then determined by $\sigma_{abs} = \sigma_{ext} - \sigma_{scat}$). Mie also introduced the extinction, absorption, and scattering efficiencies, Q_{ext}, Q_{abs}, and Q_{scat}, of the spherical nanoparticles defined as:

$$Q = \sigma/\pi R^2 \quad (1.27)$$

and that corresponds to the cross section normalized by the intersecting surface of the nanosphere (radius R).

The Mie theory allows to determine the position and the width of the plasmon resonance as well as the contribution of both absorption and scattering depending on the size, the chemical nature of the particle, and the surrounding medium nature. Small nanoparticles give rise to the distribution of dipolar charges on the surface. In the case of spherical particles, it presents one resonance mode due to homogeneity in charge displacement. The symmetry gives a resonance independent of the polarization direction of the incident light and only one plasmon band is observable.

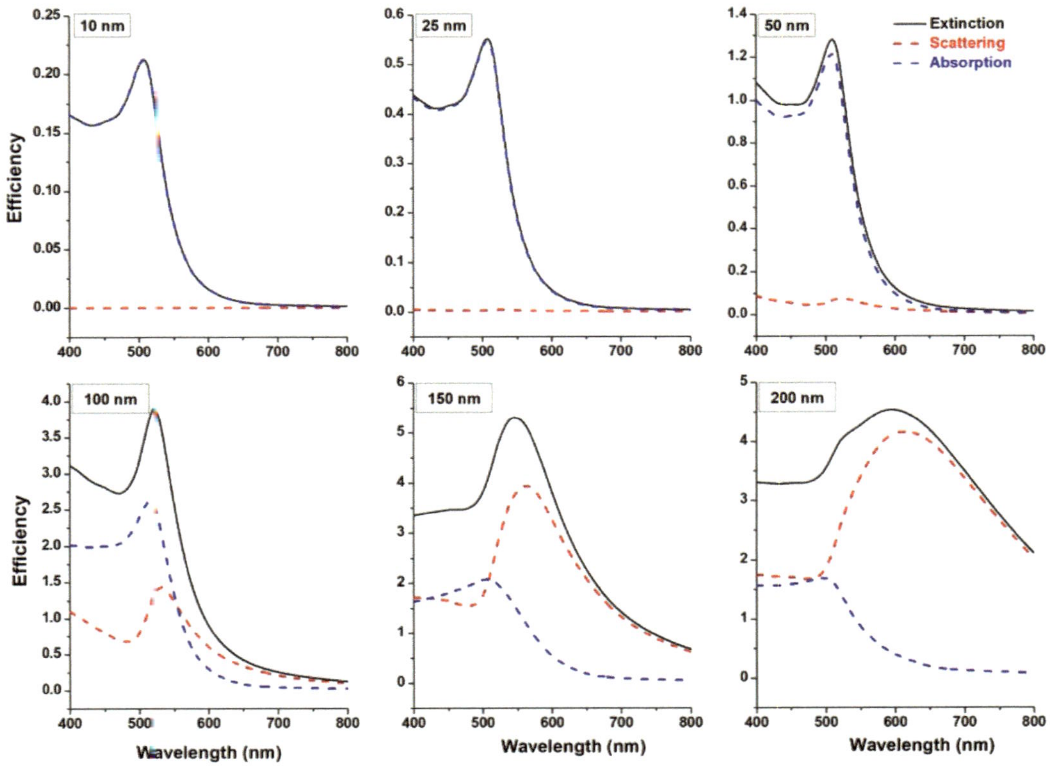

Figure 1.5 Extinction, absorption, and scattering efficiencies versus the excitation wavelength for gold nanospheres in vacuum with different diameters.

As shown in Fig. 1.5, the nanosphere size plays a role in the resonance response. First, the position of the resonance depends on the size and an increase in particle size will induce a redshift of the resonance. Second, the contribution of the absorption and the scattering to the extinction process will also depend on the size (Fig. 1.5 and Fig. 1.6). For a particle with a radius smaller than 40 nm, absorption dominates while scattering is negligible. The energy absorption in small nanoparticles is first due to the surface plasmons mode (band between 350 and 400 nm) and second due to the electron interband transitions (band lower than 300 nm).

On the contrary, in the case of nanoparticles with a bigger size, scattering dominates the resonance response whereas absorption becomes negligible [21].

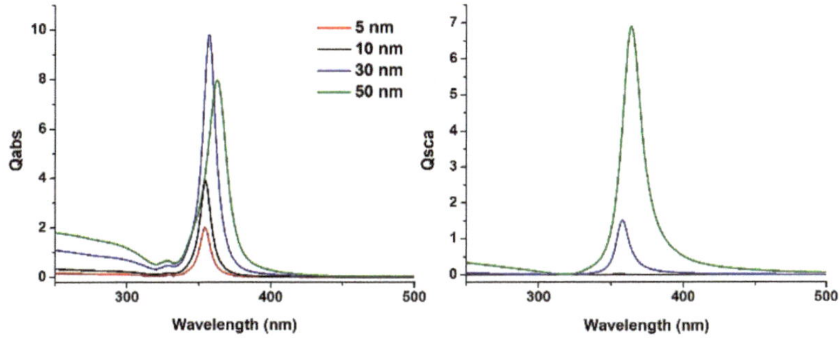

Figure 1.6 (a) Absorption and (b) scattering efficiencies versus the excitation wavelength for silver nanospheres in vacuum with radius from 5 nm to 50 nm [21].

Thus, the Mie model provides a complete description of the light–matter interaction with small particles. However, it can be tricky to use as it is necessary to calculate the whole set of Mie coefficients, which could be difficult to interpret straightly. Some simpler models can be used to approximate the nanosphere behavior and to have a better insight into the different processes that occur during the plasmon excitation as well as into the influence of the different parameters.

The first approximation that can be used is the Rayleigh or quasi-static approximation. In this way, the nanoparticle size is considered negligible compared to the wavelength ($R<<\lambda$, where R is the sphere radius), and as a consequence, the electric field amplitude is constant on the whole nanoparticle. The electron cloud is excited by the same electric field amplitude that is homogeneous at the nanoparticle scale.

The absorption and scattering cross sections and efficiencies can be estimated as:

$$\sigma_{\mathrm{Abs}} = k\mathrm{Im}(\alpha) \tag{1.28}$$

$$\sigma_{\mathrm{Scat}} = \frac{k^4}{6\pi}|\alpha|^2 \tag{1.29}$$

$$Q_{\mathrm{Abs}} = 4kR\,\mathrm{Im}\left(\frac{\varepsilon_{\mathrm{Metal}} - \varepsilon_{\mathrm{Medium}}}{\varepsilon_{\mathrm{Metal}} + 2\varepsilon_{\mathrm{Medium}}}\right) \tag{1.30}$$

$$Q_{\mathrm{Scat}} = \frac{8}{3}(kR)^4\left|\frac{\varepsilon_{\mathrm{Metal}} - \varepsilon_{\mathrm{Medium}}}{\varepsilon_{\mathrm{Metal}} + 2\varepsilon_{\mathrm{Medium}}}\right|^2 \tag{1.31}$$

where k is the light wavevector, ε_0 is the vacuum permittivity, $\varepsilon_{\mathrm{Medium}}$ is the dielectric constant of the surrounding medium, $\varepsilon_{\mathrm{Metal}}$ is the dielectric constant of the metal, R is the radius of the nanosphere, and α is the nanosphere polarizability, such as:

$$\alpha = 3V\left(\frac{\varepsilon_{\mathrm{Metal}} - \varepsilon_{\mathrm{Medium}}}{\varepsilon_{\mathrm{Metal}} + 2\varepsilon_{\mathrm{Medium}}}\right) = 4\pi R^3\left(\frac{\varepsilon_{\mathrm{Metal}} - \varepsilon_{\mathrm{Medium}}}{\varepsilon_{\mathrm{Metal}} + 2\varepsilon_{\mathrm{Medium}}}\right) \tag{1.32}$$

where V is the nanosphere volume [17].

One can notice that σ_{Abs} is proportional to the radius and inversely proportional to λ, whereas σ_{Scat} is proportional to the radius at the fourth and inversely proportional to λ at the fourth (Rayleigh scattering). For small values of R, the absorption will dominate the interaction and $Q_{\mathrm{Abs}} \approx Q_{\mathrm{Ext}}$. But for a larger nanosphere, σ_{Scat} will increase faster than σ_{Abs} and will dominate the extinction process.

The cross sections, efficiencies, and polarizability will be maximum when $\varepsilon_{\mathrm{Metal}} + 2\varepsilon_{\mathrm{Medium}} = 0$ and thus for $\varepsilon_{\mathrm{Metal}} = -2\varepsilon_{\mathrm{Medium}}$. This corresponds to the resonance condition.

It means that the nanosphere should have a negative dielectric constant (metal) and that the plasmon resonance will depend on the nature of metal and the dielectric constant of the surrounding medium. For instance, an Ag or an Au nanosphere in vacuum ($\varepsilon_{\mathrm{Metal}} = -2$) will exhibit a plasmon resonance around 350 and 500 nm.

One can notice that the resonance condition $\varepsilon_{\mathrm{Metal}} = -2\varepsilon_{\mathrm{Medium}}$ is constant and does not depend on the size or shape of the nanoparticle. The resonance position will also be constant.

It is then necessary to refine the Rayleigh approximation to have a better description of the light–nanoparticle interaction. The polarizability becomes:

$$\alpha = \frac{4}{3}\pi R^3 \left(\frac{\varepsilon_{\text{Metal}} - \varepsilon_{\text{Medium}}}{\varepsilon_{\text{Medium}} + A_{\text{eff}}\left(\varepsilon_{\text{Metal}} - \varepsilon_{\text{Medium}}\right)} \right) \tag{1.33}$$

where A_{eff} is the effective depolarization factor.

$$A_{\text{eff}} = \frac{1}{3} - \frac{1}{3}k^2R^2 - i\frac{2}{9}k^3R^3 = \frac{1}{3} - \frac{\pi V}{\lambda^2 R} - i\frac{4\pi^2 V}{3\lambda^3} \tag{1.34}$$

The first term, 1/3, is the static depolarization calculated for a sphere in the Rayleigh approximation. The two other terms are introduced to correct the Rayleigh approximation and take into account the size of the nanosphere compared to the excitation wavelength. The second term, $\frac{1}{3}k^2R^2$, is the dynamic depolarization, whereas the third term, $i\frac{2}{9}k^3R^3$, is the radiation damping [22].

The new resonance condition will then become:

$$\varepsilon_{\text{Metal}} = \varepsilon_{\text{Medium}} \left(1 - \frac{1}{A_{\text{eff}}} \right) \tag{1.35}$$

The resonance position will depend on the value of A_{eff} and thus on the particle size and more specifically on the ratio R/λ between the particle size and the excitation wavelength.

The dynamic depolarization is introduced as the nanoparticle size is no more negligible compared to the wavelength as well as the electric field amplitude is no more constant and homogeneous along the nanoparticle. When the size of the nanoparticle increases, A_{eff} decreases and the resonance condition occurs for $\varepsilon_{\text{Metal}} < -2\varepsilon_{\text{Medium}}$. The resonance position is then redshifted compared to the Rayleigh approximation. This redshift is larger for larger nanoparticles.

The radiation damping considers reduction in the energy inside the nanoparticle due to the photon radiation induced by the scattering process. As it is an imaginary term, it will reduce the resonance efficiency as well as induce a resonance redshift. The intensity of the resonance will decrease, and its width will be larger.

Using both these correction terms, we recover the effective influence of the increase in the particle size: either the redshift or the broadening of the resonance.

1.4.2 LSP in Ellipsoidal Nanoparticle

The ellipsoidal nanoparticle has no more spherical symmetry but exhibits three different axes (a, b, and c) having three different lengths. The plasmon modes are no more homogeneous whatever the polarization direction is but depend on the excited axes. Thus, in the ellipsoidal case, each axis presents one specific polarizability depending on the axes size and its aspect ratio compared to the other axes. Consequently, one ellipsoidal nanoparticle will exhibit three different dipolar modes (one mode for each axis).

In the Rayleigh approximation, the ellipsoidal polarizability for each axis is given by:

$$\alpha = V\frac{\varepsilon_{\text{Metal}} - \varepsilon_{\text{Medium}}}{\varepsilon_{\text{Metal}} + L_{\text{j}}\left(\varepsilon_{\text{Medium}} - \varepsilon_{\text{Metal}}\right)} \tag{1.36}$$

where V is the ellipsoidal nanoparticle volume ($V = \frac{4}{3}\pi abc$, where a, b, and c are the half-lengths of the axes), L_{j} is the static depolarization factor, and j is one of the three axes a, b, and c of the ellipsoid. L_{j} depends on the anisotropy of the nanoparticle and is calculated as:

$$L_{\text{j}} = \frac{1}{2}abc\int_{0}^{+\infty}\frac{dq}{\left(q+j^2\right)\sqrt{\left(q+a^2\right)+\left(q+b^2\right)\left(q+c^2\right)}} \tag{1.37}$$

and

$$L_{\text{a}} + L_{\text{b}} + L_{\text{c}} = 1 \tag{1.38}$$

In the case of nanospheres, the three L_{j}'s are equal to 1/3 due to the equivalence of each axis. The polarizability is then defined as indicated in Eq. (1.32).

For ellipsoidal nanoparticles, the L_{j} factor will have a value larger or lower than 1/3, which will modify the position of the plasmon resonance as the resonance condition is now:

$$\varepsilon_{\text{Metal}} = \varepsilon_{\text{Medium}}\left(1 - \frac{1}{L_{\text{j}}}\right) \tag{1.39}$$

When the axis length is larger than the two others, its static depolarization factor will become lower than 1/3, inducing a redshift of the plasmon resonance (it is equivalent to an increase in the axis length). In the opposite, when the axis length is lower than the two others, the static depolarization factor will become larger than 1/3, inducing a blueshift of the plasmon resonance (it is equivalent to a decrease in the axis length).

Thus, the extinction spectrum will exhibit three plasmon bands (one for each axis), blueshifted and redshifted compared to the plasmon resonance of a nanosphere having the same volume.

As for nanospheres, some corrections can be applied if the Rayleigh approximation can no more be considered and if the nanoparticle size is no more negligible compared to the excitation wavelength.

In this case, the polarizability is still determined by Eq. (1.33), but the effective depolarization factor can be written as follows [22]:

$$A_{\text{eff}} = L_{\text{j}} - \frac{1}{3}k^2 (abc)^{2/3} - i\frac{2}{9}k^3 (abc)^3 \tag{1.40}$$

The dynamic depolarization factor and the radiation damping terms are adapted to the ellipsoidal geometry. One can notice that both terms are mainly influenced by the particle volume and not by one specific shape or one axis length. Both terms have the same effect on the plasmon resonance, meaning a redshift and a broadening of the resonance when the volume of the particle increases compared to the excitation wavelength.

1.4.3 Influence of the Environment

As the LSP resonance condition depends on the dielectric constant of the surrounding medium, $\varepsilon_{\text{Medium}}$ (Eqs. (1.35) and (1.39)), the resonance position is affected by the nature of the surrounding medium and an increase in $\varepsilon_{\text{Medium}}$ will induce a redshift of the resonance. For instance, the dispersion of nanoparticles in a solution, the deposition of a thin layer, or the grafting of molecules at the nanoparticle surface will modify the position of the resonance [23, 24] as well as its width [25, 26].

Figure 1.7 shows a simulation of the extinction efficiencies, Q_{Ext}, for Ag nanoparticles embedded in different media (vacuum, water, ethanol, and glass). One can observe a redshift of the plasmon resonance for all diameters when the refractive index of the surrounding medium increases. These results demonstrate that the plasmon resonance position depends on the surrounding media [10].

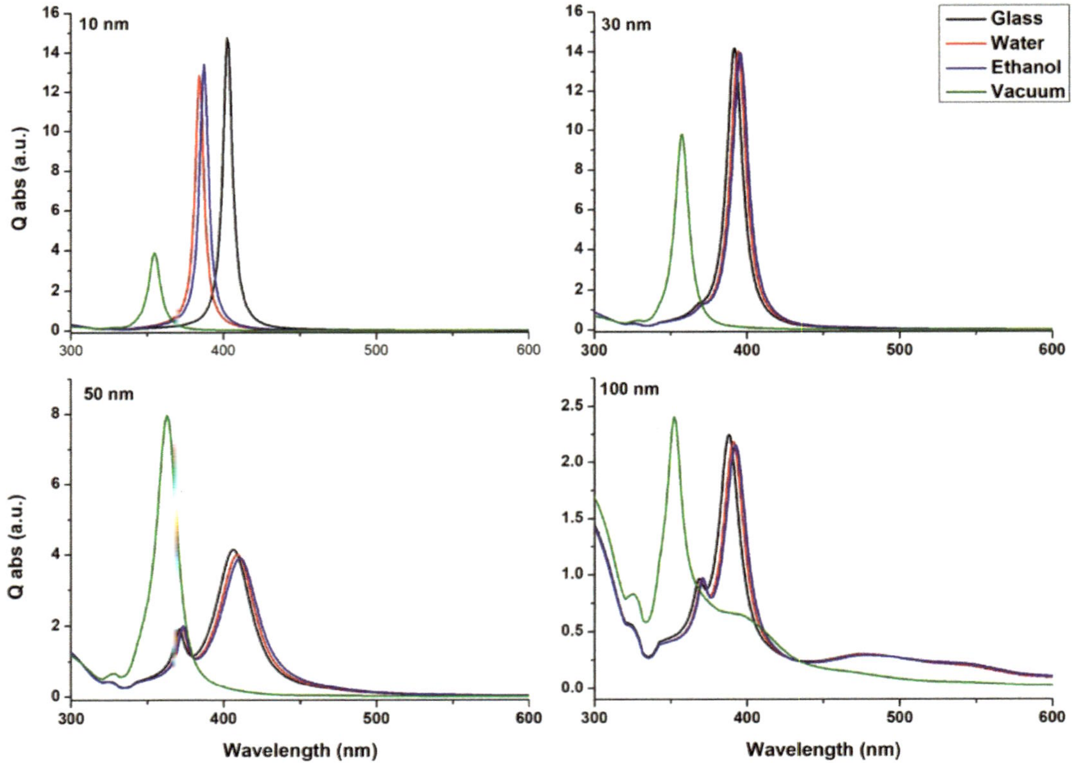

Figure 1.7 Simulation of extinction efficiencies, Q_{Ext}, for Ag nanoparticles (diameters from 10 nm to 100 nm), in different environments: vacuum (n = 1), water (n = 1.3333), ethanol (n = 1.3612), and glass (n = 1.5).

This effect is largely used in sensing as the LSPR shift can be related with the capture of a molecule at the nanoparticle surface and thus with the molecular detection [27–30]. However, the sensitivity of the LSPR sensing is largely lower than the one reached with the SPR method.

In the case of aggregation of nanoparticles in clusters, a strong electromagnetic coupling occurs between the nanoparticles when the interparticle separation is smaller than the excitation wavelength. Such plasmonic coupling induces a modification of the plasmon resonance, which can be blue- or redshifted depending on the direction of the excitation polarization [31, 32]. This nearfield interaction also induces a huge enhancement of the field compared to a single nanoparticle, creating hot spots highly localized in the nanoparticle gap [33]. The smaller the gap, the higher the field enhancement [31, 34, 35].

1.4.4 Influence of Metal Nature

In visible light, the metals commonly used to produce nanoparticles are silver (Ag), gold (Au), and recently aluminum (Al) [36, 37]. The noble metals such as Au, Ag, Cu, Pt, and Pd have their plasmon resonances in the visible range for spherical nanoparticles. They have low absorption, insuring a narrow plasmon band width and better resonance efficiency.

The real and imaginary parts of the dielectric constant for the two metals Au and Ag are given in Fig. 1.8. As their real parts are not identical, the plasmon position will change with the type of metal. For a similar nanoparticle (same size and same shape), the plasmon resonance of an Ag nanoparticle is blueshifted compared to a gold one. The width of the resonance depends on the imaginary part of the metal. Indeed, a higher imaginary part results in a higher absorption, which induces larger optical losses and larger damping of the resonance. For example, the Au absorption is minimal between 600 and 700 nm, which will result in a sharper resonance in this range [15]. Ag exhibits a lower imaginary part on the whole visible range, implying lower damping and higher field enhancement at the nanoparticle surface [38].

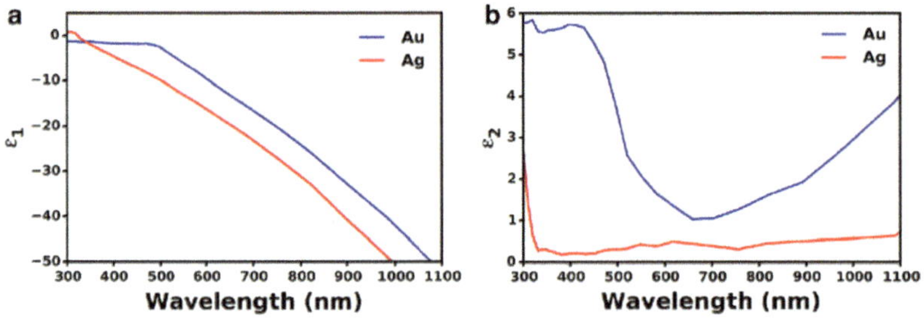

Figure 1.8 (a) Real part and (b) imaginary part of the dielectric function for Au and Ag [39].

However, Au is more stable chemically than Ag and is then used more than Ag, especially for biological applications.

For instance, Fig. 1.9 presents the simulated extinction spectra for nanospheres made of three different metals: Ag, Au, and Cu. One can observe a redshift of the plasmon resonance when the Ag metal is replaced by Au or Cu. Moreover, the plasmon resonance for Ag metal is very sharp compared to Au and Cu. This is directly related with the imaginary part of the dielectric constant, which is largely lower for Ag compared to Au and Cu. Similar experimental results have been obtained on nanocubes of Ag or Au deposited on SiN_4 (n = 2.05) or Formvar (a thermoplastic resin) (n = 1.5). A direct comparison can be done as the nanocube parameters (size, shape, and composition) were the same. As we can see, the LSPR is redshifted when Ag is replaced by Au and when the refractive index of the substrate increases [40].

1.5 Molecular Plasmonics

Plasmons can also have a large influence on their local environment. Indeed, the excitation of plasmons will induce several physical processes directly related with the excitation of the electron cloud inside nanoparticles. Thus, three processes can occur:

1. As the electron oscillation will result in charge motion inside the nanoparticle, a large electromagnetic field will be created around the nanoparticle. This induces a large field enhancement of several orders of magnitude. Thus, the nanoparticle can act as an intense nanosource of light

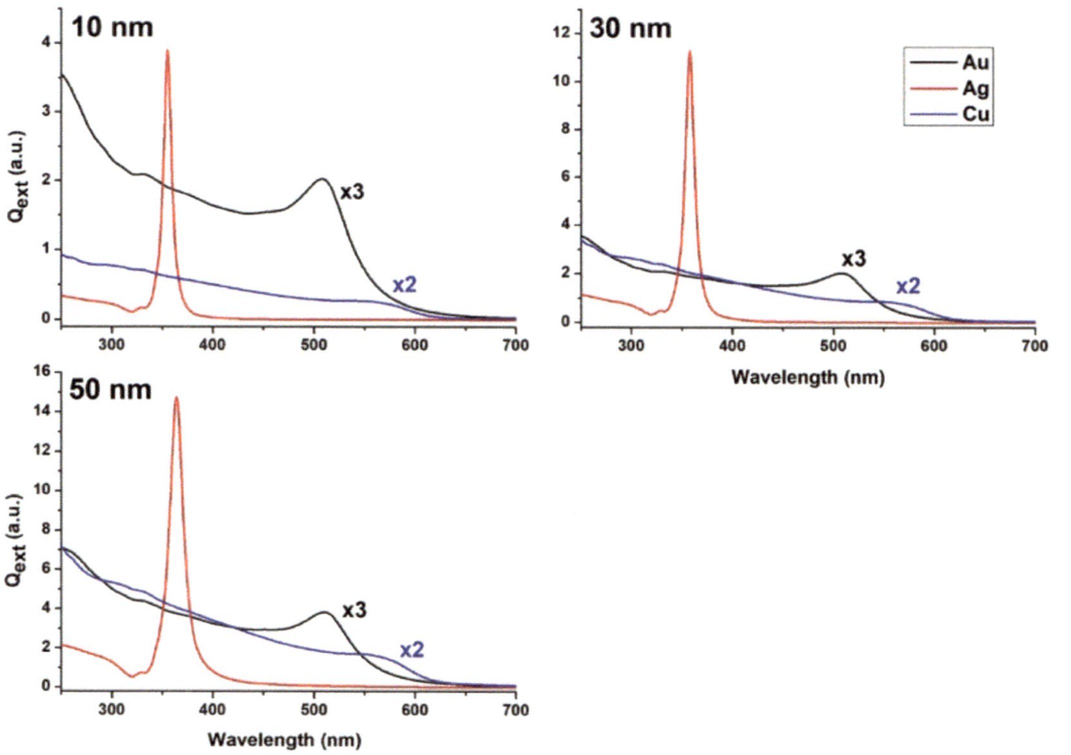

Figure 1.9 Extinction spectra of nanospheres of different diameters made of Ag, Au, and Cu in vacuum. The extinction spectra were simulated by the Mie theory.

and all photonic processes can then be enhanced by the plasmon excitation. For instance, the spectroscopical signal from molecules located at the vicinity of the nanoparticle surface can be enhanced, giving rise to all the enhanced spectroscopies such as Metal-Enhanced Fluorescence (MEF) [41], Surface-Enhanced Raman Scattering (SERS) [42–44], or Surface-Enhanced IR Absorption (SEIRA) [45–48].

2. The excitation of the plasmon induces an absorption of the photon energy by the nanoparticle. The de-excitation of the electron can occur through the phonons and induces an increase in the metal temperature, which can be transferred to the local environment. The nanoparticle then acts as a local nanoheater, and the temperature of the surrounding environment can be increased by several tens of degrees [49–53]. This effect is known as thermoplasmonics.
3. The transfer of the photon energy to the electron cloud can also produce some highly energetic electrons known as hot electrons [54, 55]. These electrons can then be transferred to molecules or materials that are in contact with or close to the nanoparticle surface.

All these processes can be exploited to modify and control the direct environment of the nanoparticle. In this context, they have already been used to induce plasmon-mediated chemical reactions as it will be detailed in the following chapters of this book [55, 56].

References

1. H. Raether (1988). *Surface Plasmon on Smooth and Rough Surfaces and on Gratings*, Springer-Verlag Berlin Heidelberg GmbH, ISBN 3-540-17363.
2. W. Fritzsche and M. Lamy de la Chapelle (2014). *Molecular Plasmonics*, Wiley-VCH, Verlag GmbH & Co., ISBN: 978-3-527-32765-2.
3. P. Drude (1900). Zur Elektronentheorie der Metalle, *Annalen der Physik*, 1, p. 566.
4. N. W. Ashcroft and N. D. Mermin (1976). *Solid State Physics*, Saunders: Philadelphia.
5. H. A. Lorentz (1905). *Proc. R. Acad. Sci. Amsterdam*, 7, p. 438.

6. M. Eldlio, F. Che, and M. Cada (2014). Drude-Lorentz model of semiconductor optical plasmons, *IAENG Transactions on Engineering Technologies*, pp. 41–49.
7. E. R. Rudberg (1930). Characteristic energy losses of electrons scattered from incandescent solids, *Proceedings of the Royal Society of London. Series A, Containing Papers of a Mathematical and Physical Character*, 127, pp. 111–140.
8. T. A. Kelf, et al. (2006). Localized and delocalized plasmons in metallic nanovoids, *Phys. Rev. B*, 74, p. 245415.
9. M. Lequeux, D. Mele, P. Venugopalan, R. Gillibert, S. Boujday, W. Knoll, J. Dostalek, and M. Lamy de la Chapelle (2020). Plasmonic properties of gold nanostructures on gold film, *Plasmonics*, 15, p. 1653.
10. A. Otto (1968). Excitation of nonradiative surface plasma waves in silver by the method of frustrated total reflection, *Z. Physik*, 216, pp. 398–410.
11. E. Kretschmann (1971). Die Bestimmung optischer Konstanten von Metallen durch Anregung von Oberflächenplasmaschwingungen, *Zeitschrift fur Physik*, 241, pp. 313–324.
12. D. Barchiesi, D. Macías, L. Belmar-Letellier, D. van Labeke, M. Lamy de la Chapelle, T. Toury, E. Kremer, L. Moreau, and T. Grosges (2008). Plasmonics: influence of the intermediate (or stick) layer on the efficiency of sensors, *Applied. Phys. B-Lasers and Optics*, 93, p. 177.
13. D. Barchiesi, N. Lidgi-Guigui, and M. Lamy de la Chapelle (20012). Functionalization layer influence on the sensitivity of surface plasmon resonance (SPR) biosensor, *Optics Communications*, 285(6), p. 1619.
14. J. Grand, P.-M. Adam, A.-S. Grimault, A. Vial, M. Lamy de la Chapelle, J.-L. Bijeon, S. Kostcheev, and P. Royer (2006). Optical extinction spectroscopy of oblate, prolate and ellipsoid shaped gold nanoparticles:experiments and theory, *Plasmonics*, 1(2-4), p. 135.
15. F. J. Colas, et al. (2016). Red-shift effects in surface enhanced raman spectroscopy: spectral or intensity dependence of the near-field?, *The Journal of Physical Chemistry C*, 120, pp. 13675–13683.
16. G. Mie (1908). *Ann. Physik*, 25, p. 377.
17. C. F. Bohren and D. R. Huffman (2004). *Absorption and Scattering of Light by Small Particles*, Wiley-VCH, Verlag GmbH & Co., ISBN: 978-0-471-29340-8.
18. F. Papoff and B. Hourahine (2011). Geometrical Mie theory for resonances in nanoparticles of any shape, *Opt. Express, OE*, 19, pp. 21432–21444.

19. A. Amirjani, F. Firouzi, and D. F. Haghshenas (2020). Predicting the size of silver nanoparticles from their optical properties, *Plasmonics*, 15, pp. 1077–1082.
20. J. S. Duque, J. S. Blandón, and H. Riascos (2017). Localized plasmon resonance in metal nanoparticles using Mie theory, *J. Phys.: Conf. Ser.*, 850, p. 012017.
21. C. Noguez (2005). Optical properties of isolated and supported metal nanoparticles, *Optical Materials*, 27, pp. 1204–1211.
22. A. Wokaun (1985). Surface enhancement of optical fields mechanism and applications, *Molecular Physics*, 56(1), pp. 1–33.
23. L. Liz-Marzán (2020). *Colloidal Synthesis of Plasmonic Nanometals*, 1st Ed. (New York).
24. R. Gillibert, M. Sarkar, J. Moreau, M. Besbes, M. Canva, and M. L. de la Chapelle (2016). Near-field enhancement localization on plasmonic gratings. *The Journal of Physical Chemistry C*, 120, pp. 27562–27570.
25. M. Lamy de la Chapelle, H. Shen, N. Guillot, B. Fremaux, B. Guelorget, and T. Toury (2013). New gold nanoparticles adhesion process opening the way of improved and highly sensitive plasmonics technologies, *Plasmonics*, 8(2), p. 411.
26. F. Colas, D. Barchiesi, S. Kessentini, T. Toury, and M. Lamy de la Chapelle (2015). Comparison of adhesion layers of gold on silicate glasses for SERS detection, *Journal of Optics*, 17, p. 114010.
27. G. Barbillon, J. L. Bijeon, J. Plain, M. Lamy de la Chapelle, P. M. Adam, and P. Royer (2007). Biological and chemical gold nanosensors based on localized surface plasmon resonance, *Gold Bulletin*, 40(3), p. 240.
28. G. Barbillon, J. L. Bijeon, J. Plain, M. Lamy de la Chapelle, P. M. Adam, and P. Royer (2007). Electron beam lithography designed chemical nanosensors based on localized surface plasmon resonance, *Surface Science*, 601(21), p. 5057.
29. G. Barbillon, J.-L. Bijeon, J.-S. Bouillard, J. Plain, M. Lamy de la Chapelle, P.-M. Adam, and P. Royer (2008). Detection in near-field domain of biomolecules adsorbed on a single metallic nanoparticle, *Journal of Microscopy*, 229(3), p. 385.
30. J. Huang, Y. Zhang, Z. Lin, W. Liu, X. Chen, Y. Liu, H. Tian, Q. Liu, R. Gillibert, J. Spadavecchia, N. Djaker, M. Lamy de la Chapelle, Y. Xiang, and W. Fu (2019). Femtomolar detection of nucleic acid based on functionalized gold nanoparticles, *Nanophotonics*, 8(9), p. 1495.
31. P. K. Jainab and M. A. El-Sayed (2010). Plasmonic coupling in noble metal nanostructures, *Chemical Physics Letters*, 487, pp. 153–164.
32. W. Rechberger, et al. (2003). Optical properties of two interacting gold nanoparticles, *Optics Communications*, 220, pp. 137–141.

33. M. Pelton, J. Aizpurua, and G. Bryant (2008). Metal-nanoparticle plasmonics, *Laser & Photonics Reviews*, 2, pp. 136–159.

34. X. Huang, I. H. El-Sayed, W. Qian, and M. A. El-Sayed (2007). Cancer cells assemble and align gold nanorods conjugated to antibodies to produce highly enhanced, sharp, and polarized surface Raman spectra: a potential cancer diagnostic marker, *Nano Letters*, 7, pp. 1591–1597.

35. S. Kessentini, D. Barchiesi, C. D'Andrea, A. Toma, N. Guillot, E. Di Fabrizio, B. Fazio, O. Marago, P. Gucciardi, and M. Lamy de la Chapelle (2014). Gold dimer nanoantenna with slanted gap for tunable LSPR and improved SERS, *J. Phys. Chem. C*, 118(6), p. 3209.

36. A. Muravitskaya, et al. (2020). Refractive index mediated plasmon hybridization in an array of aluminium nanoparticles, *Nanoscale*, 12, pp. 6394–6402.

37. R. Gillibert, F. Colas, R. Yasukuni, G. Picardi, and M. Lamy de la Chapelle (2017) Plasmonic properties of aluminium nanocylinders in the visible range, *J. Phys. Chem. C*, 121(4), p. 2402.

38. W. R. Erwin, H. F. Zarick, E. M. Talbert, and R. Bardhan (2016). Light trapping in mesoporous solar cells with plasmonic nanostructures, *Energy & Environmental Science*, 9, pp. 1577–1601.

39 P. B. Johnson and R. W. Christy (1972). Optical constants of noble metals, *Phys. Rev. B*, 6, p. 4370.

40. E. Ringe, et al. (2010). Unraveling the effects of size, composition, and substrate on the localized surface plasmon resonance frequencies of gold and silver nanocubes: a systematic single-particle approach, *ACS Publications*, 114(29), pp. 12511–12516.

41. S. Boujday, M. Lamy de la Chapelle, J. Srajer, and W. Knoll (2015). Enhanced vibrational spectroscopies as tools for small molecule biosensing, *Sensors*, 15, p. 21239.

42. C. D. Geddes and J. R. Lakowicz (2002). Metal-enhanced fluorescence, *J. Fluores.*, 12, p. 121.

43. N. Guillot and M. Lamy de la Chapelle (2012). The electromagnetic effect in surface enhanced Raman scattering: enhancement optimization using precisely controlled nanostructures, *Journal of Quantitative Spectroscopy and Radiative Transfer*, 113, p. 2321.

44. S. Bell, G. Charron, E. Cortes, J. Kneipp, M. Lamy de la Chapelle, J. Langer, M. Prochazka, V. Tran, and S. Schlücker (2020). Towards reliable and quantitative SERS: from key parameters to good analytical practice, *Angew. Chem. Int. Ed.*, 59, p. 5454.

45. F. Neubrech, A. Pucci, T. Cornelius, S. Karim, A. García-Etxarri, and J. Aizpurua (2008). *Phys. Rev. Lett.* 101, p. 157403.

46. A. Pucci, F. Neubrech, D. Weber, H. Shen, T. Toury, and M. Lamy de la Chapelle (2010). Surface enhanced infrared spectroscopy using gold nanoantennas, *Physica Status Solidi B*, 247(8), p. 2071.

47. C. D'Andrea, J. Bochterle, A. Toma, C. Huck, F. Neubrech, E. Messina, B. Fazio, O. M. Marago, E. Di Fabrizio, M. Lamy de la Chapelle, P. G. Gucciardi, and A. Pucci (2013). Optical nanoantennas for multiband surface-enhanced infrared and Raman spectroscopy, *ACS Nano*, 7(4), p. 3522.

48. F. Verger, F. Colas, O. Sire, H. Shen, E. Rinnert, K. Boukerma, V. Nazabal, C. Boussard-Plédel, B. Bureau, T. Toury, M. Lamy de la Chapelle, and C. Compère (2015). Surface enhanced IR absorption by nanoantenna on chalcogenide glass substrates, *Applied Physics Letter*, 106(7), p. 073103.

49. G. Baffou, P. Berto, E. Bermudez Urena, R. Quidant, S. Monneret, J. Polleux, and H. Rigneault (2013). Photoinduced heating of nanoparticle arrays, *ACS Nano*, 7(8), pp. 6478–6488.

50. G. Baffou and R. Quidant (2013). Thermo-plasmonics: using metallic nanostructures as nano-sources of heat, *Laser Photonics Rev.*, 7(2), pp. 171–187.

51. G. Picardi, F. Colas, R. Gillibert, and M. Lamy de la Chapelle (2016). Spectral shift of the plasmon resonance between the optical extinction and absorption of gold and aluminum nanodisks, *J. Phys. Chem. C*, 120(45), p. 26025.

52. H. Moustaoui, J. Saber, I. Djeddi, Q. Q. Liu, A. Diallo, J. Spadavecchia, M. Lamy de la Chapelle, and N. Djaker (2019). Shape and size effect on photothermal heat elevation of gold nanoparticles: absorption coefficient experimental measurement of spherical and urchin-shaped gold nanoparticles, *J. Phys. Chem. C*, 123(28), p. 17548.

53. R. Gillibert, F. Colas, M. Lamy de la Chapelle, and P. Gucciardi (2020). Heat dissipation of metal nanoparticles in the dipole approximation, *Plasmonics*, 15, p. 1001.

54. A. Giugni, B. Torre, A. Toma, et al. (2013). Hot-electron nanoscopy using adiabatic compression of surface plasmons, *Nature Nanotech*, 8, pp. 845–852.

55. G. Baffou and R. Quidant (2014). Nanoplasmonics for chemistry, *Chem. Soc. Rev.*, 43, p. 3898.

56. E. Cortés, L. V. Besteiro, A. Alabastri, A. Baldi, G. Tagliabue, A. Demetriadou, and P. Narang (2020). Challenges in plasmonic catalysis, *ACS Nano*, 14(12), pp. 16202–16219.

Chapter 2

Characterization of Plasmonic Nanomaterial

Gennaro Picardi,[a,b] Mathieu Edely,[a] and Marc Lamy de la Chapelle[a]
[a]*Le Mans University, Laboratoire IMMM, UMR6283, Avenue Olivier Messiaen 72085 Le Mans Cedex 9, France*
[b]*Nantes University, CNRS, CEISAM, UMR 6230, F-44000 Nantes, France*
marc.lamydelachapelle@univ-lemans.fr

As the plasmonic properties of metallic nanoparticles and nanostructures are related to their size and shape, experimental investigations to determine their optical and structural properties have been implemented. These techniques have been applied to study the morphology and structure of the plasmonic materials at the sub-nanometer scale.

2.1 Ultraviolet–Visible Spectroscopy

Ultraviolet–visible (UV–vis) spectroscopy is used to obtain the extinction spectra of a compound in solution or solid form.

Plasmonic Nanomaterials: Characterization and Applications in Organic Synthesis and Catalysis
Edited by Clémence Queffélec and D. Andrew Knight

ISBN 978-981-5129-09-0 (Hardcover), 978-1-003-47406-7 (eBook)
www.jennystanford.com

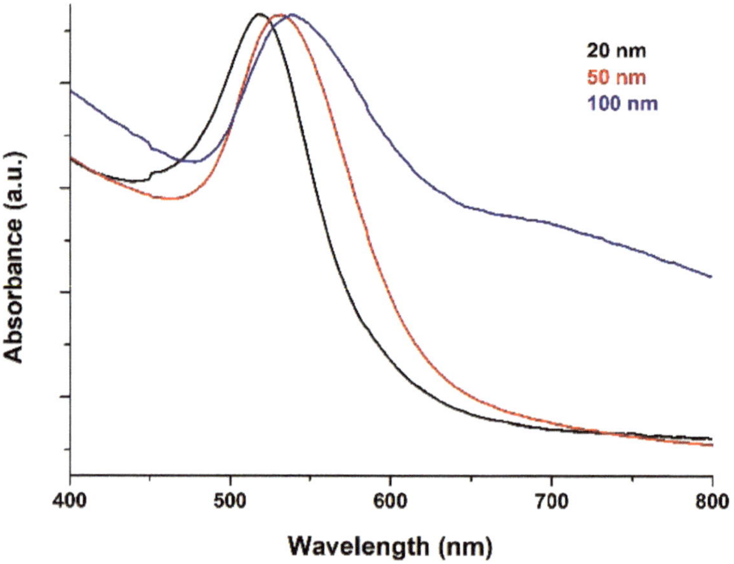

Figure 2.1 Extinction spectra (absorbance = $-\log_{10}$(transmittance)) measured for gold nanospheres in water for three different diameters: 20 nm (black), 50 nm (red), and 100 nm (blue).

Often called absorption spectroscopy because of the ability to measure the concentration of an absorbing molecule in solution, thanks to the Beer–Lambert law, it is rather extinction spectroscopy. For liquid compounds, samples are placed in a cuvette and a UV–vis spectrometer measures transmission over a fixed path length (typically 1 cm). The compound receives an incident radiation of intensity I_0, and it scatters a part of it and absorbs the other one. The transmitted intensity (I_t) of the radiation from the cell is, therefore, lower than the intensity of the initial radiation. It corresponds to the incident intensity reduced by the light scattering and absorption induced during the interaction with the nanoparticles. Extinction or transmittance is defined as $-\log(I_t/I_o)$. As presented in Chapter 1, the optical extinction is the sum of scattering and absorption. Considering a solution containing metallic nanoparticles, these two processes will be enhanced when the LSPs are excited upon illumination. The relative efficiencies of absorption and scattering will depend on the size and shape of the nanoparticles. Absorption is dominant for smaller nanoparticles, and scattering dominates for larger ones. Although the extinction spectrum cannot provide supplementary information that can be given by absorption and scattering (higher-order resonances notably sensitive to roughness or faceting of NPs) [1], it shows a broad band when the plasmon is excited. The maximum of this band corresponds to the plasmonic resonance (Fig. 2.1).

2.2 Scanning Electron Microscopy and Transmission Electron Microscopy

Scanning electron microscopy (SEM) and transmission electron microscopy (TEM) allow a drastic reduction in the micrometric resolution of optical microscopy by using electron beams with wavelengths considerably smaller than those of visible light.

2.2.1 Scanning Electron Microscopy

Scanning electron microscopy (SEM) is a technique for observing the topography of surfaces. It is mainly based on the detection of secondary electrons originating from the sample after the impact of a very fine beam of primary electrons that scans the observed surface. This detection allows to obtain images with a lateral accuracy between 1 and 10 nm depending on the source of electrons—an electron gun kept under vacuum. The gun has a tip from which electrons are produced. The electrons are extracted from the apex of the tip using a high voltage and then accelerated.

When the beam penetrates the sample, various interactions occur. Inelastic ones induce a progressive loss of energy of the electrons through interactions with the atoms of the material (electrons and nuclei). Elastic ones provoke abrupt variations in the direction of the incident electrons. These interactions between the incident electrons and the sample lead to the emission of three types of electrons from the surface: (a) backscattered electrons made of primary electrons, which are in collision with the nuclei and electrons; (b) secondary electrons, which are slowed-down primary electrons; and (c) valence electrons, which are removed by ionization, and Auger electrons made of de-excited electrons after ionization, which are characteristics of the probed chemical element. Other phenomena occur: X photons are emitted after the ionization of atoms by the incident electrons, and the emission of electromagnetic waves can appear in the case of insulted or semiconductor samples. These emission processes depend on the energy of the incident electrons, the nature of the sample (atoms and their atomic number), and the angle between the incident beam and the sample surface. The SEM image is formed from the signals of the different detectors used to collect mostly the backscattered and secondary electrons. Backscattered electrons are mainly sensitive to the surface topography and the chemical composition of the sample and come from a deep and large area (hundreds of nanometers). Secondary electrons originate from a small area around the impact of the electron beam. The lateral accuracy of the SEM depends on the electron source, whose efficiency and performance are defined by its brilliance, which is linked to the current density. Different technologies have been developed to increase brightness. The most classical one is the tungsten filament, which emits electrons by the thermoelectric effect when heated (around 2800 K). With this kind of source, the lateral accuracy is between 5 and 10 nm. The brilliance can be improved by the addition of lanthanum hexaboride (LaB_6) crystals on the tungsten filament extremity. In this case, the accuracy can reach 5 nm. The brilliance of the source can be drastically improved by using a field emission gun (FEG) that amplifies the electron extraction. A cold FEG and a Schottky effect gun can thus reach a lateral accuracy of around 1 nm.

2.2.2 Transmission Electron Microscopy

Transmission electron microscopy (TEM) focuses on the electrons that not only penetrate the sample but also pass through it. While SEM works with electrons accelerated between 1 and 30 keV, the electrons in TEM undergo acceleration at voltages between 30 and 300 keV. Conventional TEM can be equipped with a scanning mode. Scanning transmission electron microscopy (STEM) can access other signals that cannot be spatially correlated in TEM. Elastically scattered electrons are exploited to build images by either phase or diffraction contrast. For phase contrast, the image is derived from the interferences between transmitted, scattered, and diffracted beams. The diffraction contrast originates from the local variations in diffraction conditions. Inelastically scattered electrons can be used to obtain electron energy loss spectra (see later).

Thanks to their accuracies, SEM and TEM have been widely used to characterize the numerous plasmonic nanostructures developed to promote surface-enhanced Raman scattering (SERS). The colloids made of one or several materials with different shapes and sizes have been observed with both SEM and TEM after immobilization on transparent films or grids: A drop of colloids is deposited on the chosen support and evaporated at room temperature. Figure 2.2 shows a small overview of SEM and TEM images of such nanostructures.

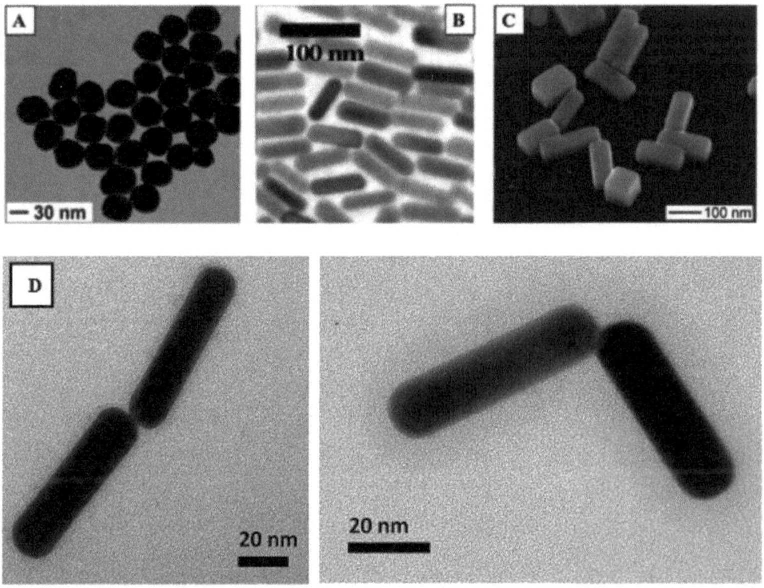

Figure 2.2 (A) SEM image of spherical gold nanoparticles; (B) SEM image of gold nanorods; (C) SEM image of silver nanobars; (D) TEM image of gold nanorods dimer. Adapted from Refs. [2, 3].

The metallic nanostructures fabricated on a substrate by lithographic techniques such as nanosphere lithography (NSL), electron beam lithography (EBL), nanoimprint lithography (NIL), or laser interference lithography (LIL) have been directly observed with SEM [4–7]. Figure 2.3 shows SEM images of SERS substrate fabricated with EBL.

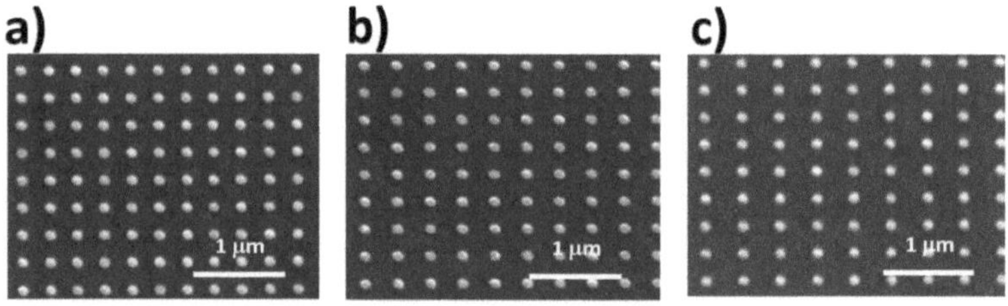

Figure 2.3 Gratings of 150 nm diameter gold nanocylinders having the same edge-to-edge period (200 nm) in one direction and different values of (a) 200 nm, (b) 250 nm, and (c) 300 nm in the other direction.

Thanks to these observations, the research groups managed to demonstrate the influence of the morphological features (size, shape, grating period) of the nanostructures thus produced on the plasmonic properties and on the SERS efficiency.

2.3 Energy-Dispersive X-Ray Spectroscopy

Among the non-electronic signals coming from the sample after irradiation with the electron beam, the X photons can be exploited for chemical analysis. These photons are the result of ionization of the elements of the sample by the primary electrons. The electron hole thus created leads to a rearrangement that can occur either by a radiative emission of X photons or by the emission of an Auger electron. Inside the chamber, the X detector allows to measure the X-ray spectrum characteristic of the elements that constitute the sample. Using this signal, it is possible to identify the chemical elements constituting the surface and to provide a chemical characterization of the material.

2.4 Electron Energy Loss Spectroscopy

Electron energy loss spectroscopy (EELS) is part of surface analysis techniques. It is mainly used to study the elemental composition and electronic structure of materials. Fast electrons of chosen energy irradiate the surface and are slowed down due to the interactions with the matter. EELS measures this electron energy loss, which is mainly due to the interaction with (a) the collective excitation of phonons (vibration of ion network) and plasmons (valence electrons) and (b) molecules adsorbed on the surface. EELS is performed in STEM, including a dispersive analyzer that separates the electrons as a function of their energy. The electron beam scans the sample, and at a fixed position, the EELS spectrum is recorded. The EELS spectrum is constituted of several peaks at different energies corresponding to the phenomena described earlier. The most intense peak deals with electrons that have not been slowed down. For energy losses between 100 eV and a few meV, the EELS spectrum concerns optical phenomena. For energy losses higher than 100 eV, transitions related to elemental composition are involved. A spectral image can be constructed from the spectral data acquired for each sample position. EELS has been used to observe

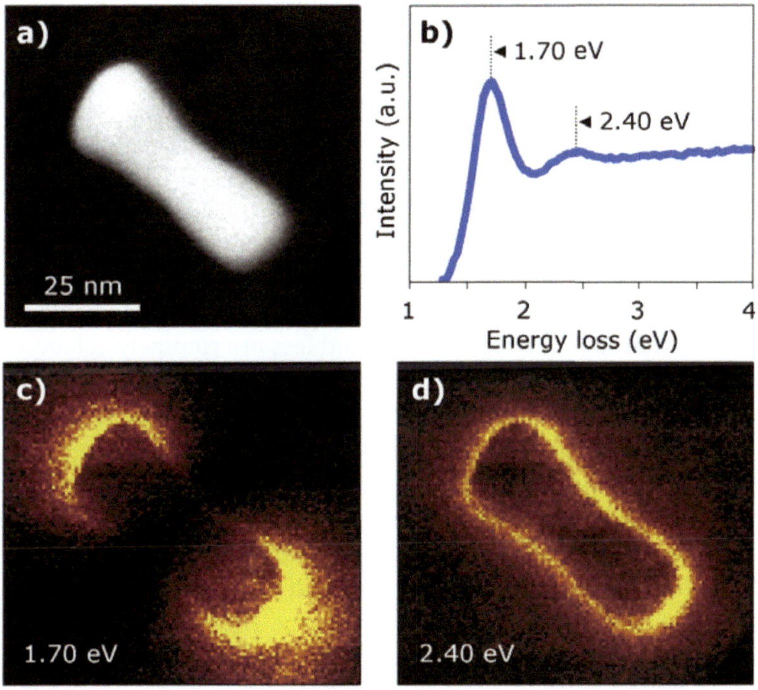

Figure 2.4 (a) STEM image of gold nanorod (annular dark field), (b) EEL spectrum, (c, d) maps of the intensity of the plasmon resonance at 1.70 eV (c) and 2.40 eV (d). Adapted from Ref. [9].

the energy and distribution of surface plasmons by comparing the changes in spectral information with the shape of nanostructures given by TEM images. Bosman et al. managed to map the distribution of surface plasmon resonance of gold nanorods and silver ellipsoids [8]. The surface plasmons were excited by electrons of high energy, 100 keV. The maps were acquired by scanning the sample, and for each pixel grid, an EEL spectrum is extracted. Spectrum images are then analyzed with principal component analysis (PCA). Figure 2.4 shows the comparison of STEM images and the intensity mapping of gold nanorods.

2.5 Atomic Force Microscopy

The first atomic force microscope (AFM) was realized in 1986 by Binning, Quake, and Gerber; its precursor, the scanning tunneling microscope (STM), granted the same year the Nobel Prize in Physics to its inventors, Binning and Rohner. Both these techniques and other related ones are commonly grouped as scanning probe microscopies (SPMs). Unlike more familiar and preceding optical microscopies, SPMs do not use light to observe the matter but rather a sharp, solid probe to touch, feel the sample surface. These non-optical microscopies are mostly used to probe the surface topography, and their foremost asset is the ability to accurately image a surface with a resolution much beyond the diffraction limit of visible light. Resolutions of a few nanometers and less are routinely achieved by SPMs.

AFM makes use of the short-distance forces that locally arise between the atoms on the sample surface and the atoms located at the very end of a very sharp probe tip. The tip probes these interaction forces. These are believed to be an ensemble of forces, including van der Waals interactions, capillary forces, and repulsive forces. All of these are only effective at separations of less than a few nanometers and rise very quickly as the separation is diminished (Fig. 2.5). Hence, a very high spatial resolution and exceptional sensitivity to even small changes in the tip to sample separation are witnessed. To explore areas as small as a few square nanometers, the scanning process involves continuously moving the tip over the surface, scanning line by line, due to the typically tiny size of the tip apex, which is approximately 10 nm. During each line scan, the height of surface features is measured by monitoring the local forces as the tip sample separation changes. The software correlates the actual *xy* position of the tip with the vertical distance of the signal and reconstructs a 2D image from the totality of the scan lines.

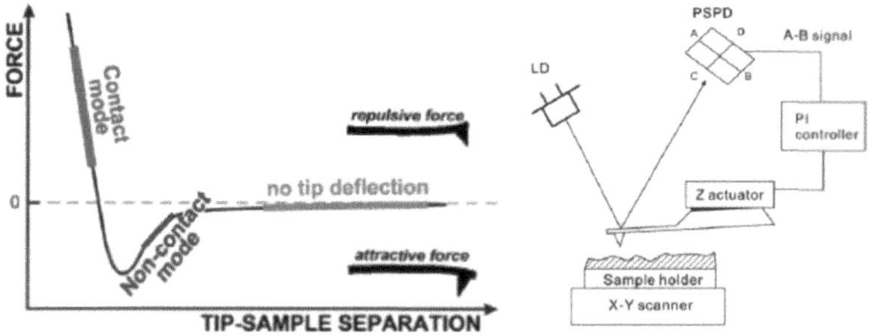

Figure 2.5 (Left) Interatomic forces acting between tip and sample as a function of their separation. Two distinct regions are defined: Negative forces are attractive and pass through a minimum—the region of pure non-contact is evident; positive forces are repulsive and rise very rapidly. Intermittent contact encompasses both attractive and repulsive forces. Physical contact begins at zero force. (Right) Generic diagram of an AFM. The laser beam from a small laser diode (LD) is focused on the very end of the cantilever and oriented so that the beam reflected from the back side of the cantilever falls within the responsive region of a PSPD. The sample is mounted on an XY scanning stage. The vertical movement of the cantilever is enabled by the Z piezo element. The photodiode signal is fed into the controlling electronics, which regulates the feedback loop and controls the three-axis movement of the tip with respect to the sample. Adapted from Refs. [10, 11].

There are three possible AFM operation modes—contact mode, tapping mode, and non-contact mode—mainly differing in the way the atomic forces are probed and which signal is exploited to render the height of the surface features. In the contact mode, the tip is mounted at the end of a supple cantilever (see later and Fig. 2.7). The cantilever is lowered toward the surface until the tip touches the "surface," which means until atomic forces set up between tip and surface. First, the tip is attracted toward the surface (longer range, weak capillary forces, see Fig. 2.5), then repulsive forces become predominant at shorter separations and rapidly increase beyond the contact point as the tip presses on the surface. Before any interaction, the cantilever is perfectly flat, but it then bends downward or upward as the tip is attracted or repulsed from the surface. This bending of the cantilever (optically detected as described later) is used as the signal monitored during the tip scanning (Fig. 2.6). In the usual initial procedure, a target value for the force to be applied by the tip is chosen by the operator (usually a small positive force, which means a certain amount of upward bending of the cantilever) and the tip starts its descent toward the surface; the approach is stopped when the target force is measured: the tip is said to be in contact. When the tip scans the surface, the AFM electronics operate to maintain this force (i.e., the cantilever deflection) constant at all the time. The difference between the target value and the actual signal during scanning is called the error signal, and the AFM electronics strive to keep this at zero all the time (technically, it is said that the electronics maintain a feedback loop). If the tip encounters a protrusion on the surface, the tip will need to climb to keep the force constant (since this depends on the separation tip to surface) and the cantilever will bend more upward. On the opposite, when encountering a depression, the cantilever will shortly bend downward looking for the missing surface to which the target force is applied until the electronics readjust the separation tip to surface.

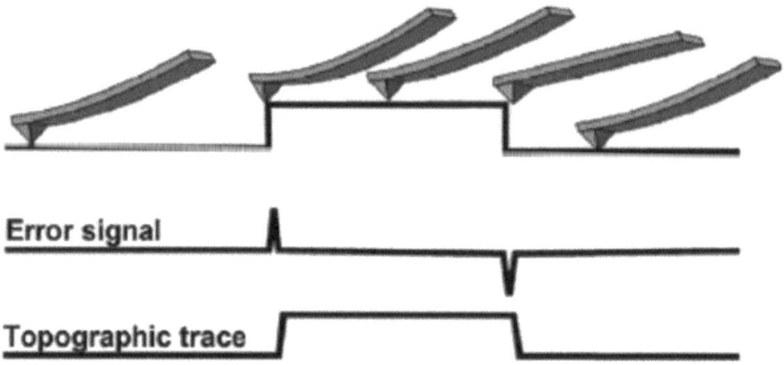

Figure 2.6 Schematic representation of the formation of the topographic trace as the tip scans the surface in the contact mode. The scanning sense is supposed to be from left to right. The error signal is constant when the tip surfs over a smooth area, and a spike results when the tip suddenly climbs (or descends) the step. Note that in the drawing, the bending of the cantilever has changed in relation to these points. Adapted from Ref. [10].

In the tapping (or intermittent) mode, a much stiffer and longer cantilever is made to oscillate at very high frequency above the sample. The oscillations are driven by a tethering small piezoelectric element operating at or near the mechanical resonance frequency of the cantilever. The oscillation amplitude can be up to 100 nm, more typically a few tens of nanometers. As the tip approaches the surface, its motion is affected by the whole range of atomic forces, first attractive and then repulsive, and at each cycle, the tip touches, taps the surface applying a force to it, even if for a short time (intermittent contact). An oscillation amplitude slightly lower than the value resulting when the tip is far away from the surface ("freely oscillating") is set as the target value and used by the electronics to define the tip to sample distance, in this case an average distance over many cycles. This operation mode of the feedback loop is called amplitude modulation. Other than the amplitude, interaction of the tip with the sample also modifies the oscillation frequency and its phase. It is then possible to operate in the tapping mode under frequency modulation, in which case a given (positive or negative) shift in the resonance frequency from the freely oscillating value is defined as the target value for the electronics.

The non-contact mode is very similar to the tapping mode and was developed before it by Wickramasinghe et al. in 1987 [11]. The main difference is a smaller amplitude oscillation (nm or less), restricting the interactions only to the range of attractive forces. Clearly, the non-contact mode is supposed to be gentler to the surface (and tip as well) than the contact or intermittent mode and is better suited to non-invasively probe soft surfaces, like biological objects. In many commercial instrumentation, it is not obvious to discriminate between operation in the tapping mode and the true non-contact mode or this is left to the expertise of the user. The frequency modulation approach allows to better discriminate between these somewhat overlapping regimes and is normally preferred.

We will briefly discuss some important components common to most AFM: the tips, the scanning elements, and the detection system connected to the controlling electronics. AFM tips are batch fabricated by a combination of lithographic and etching techniques starting from a 4" or 6" Si(100) wafer. Up to ~400 identical tips can be carved out on a 4" wafer. Also, an SiN thin layer can be introduced during the fabrication to obtain tips with different mechanical properties. The tip itself is located at the end of a triangular or rectangular cantilever (length 50 to 500 μm, width 20 to 50 μm, thickness 0.4 to 8 μm), which extends from a macroscopic holder (called chip, approximately 2 mm × 4 mm) for easy handling and mounting of the fragile cantilever (Fig. 2.7). The tip has most often a pyramidal shape (3 to 20 μm height) with the final radius of curvature < 15 nm.

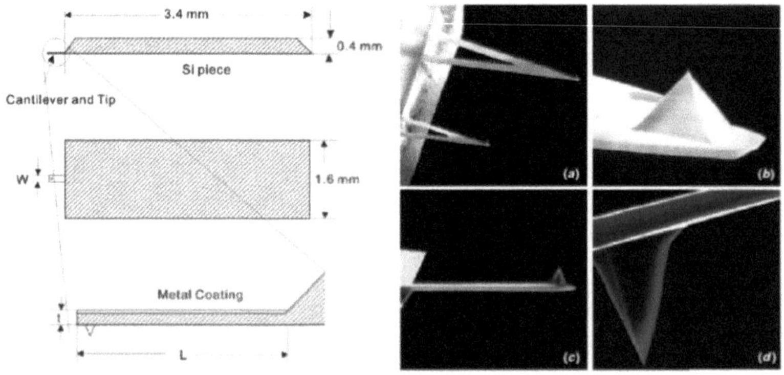

Figure 2.7 (Left) Example of an AFM probe showing the larger chip, the rectangular cantilever, and the tip at its end. In this example, a thin metal layer has been added by evaporation to the back side of the cantilever to improve the laser diode reflection and performance of the optical detection. Adapted from Ref. [13]. (Right) A gallery of SEM images of AFM cantilevers and tips: (a) two different triangular cantilevers supported by the same chip (the two cantilevers have different force constants), (b) an SiN tip at the end of a triangular cantilever, (c) an Si rectangular cantilever, and (d) an Si tip at the end of a rectangular cantilever. Adapted from Refs. [12, 13].

Different tips vary in the exact geometrical shape or the positioning, orientation, and tilting of the tip with respect to the flat cantilever and have been developed for specific applications. Special tips for high-resolution imaging have a smaller radius of only a few nanometers or present a narrower aperture (i.e., higher aspect ratio) in the final portion to probe deeper and narrower surface features, which would be inaccessible to larger tips. A conductive coating is added when the AFM is used to measure the electrical properties of a surface, such as the surface conductivity in conductive AFM and the surface potential in Kelvin probe AFM. More sophisticated functionalization of the original Si/SiO_2 surface has been developed for biological application (note that a 1–2 nm thick native oxide spontaneously and quickly forms on any exposed Si surface). Apart from the physical dimensions, a tip (or more precisely the whole cantilever carrying the tip) is characterized by its force constant in N/m and resonant frequency. As a rule of thumb, tapping-mode AFM employs very rigid cantilevers with high resonant frequencies (typically ~300 kHz, ~40 N/m), while for contact-mode AFM, the cantilever must be supple and easily bend even for small applied forces (force constant ranging from 1 to 0.05 N/m). The operation of tapping mode in liquid is possible using a cantilever with lower resonant frequencies. The performance of the AFM depends very much on the proper choice of the tip best suited to image a particular surface. Hard, clean surfaces are the easiest to image, while biological objects are more challenging. Also, AFM works best when the height gradients are not too abrupt, which usually leads to tip damaging since the tip does not have sufficient time to properly follow the surface.

The precise movement of the tip with respect to the sample surface is controlled by piezo elements. These are piezoelectric ceramics, which can elongate (compress) upon the application of an external voltage. In most of the modern AFMs, the XY and Z movements are decoupled: one piezo element (Z actuator) controls the tip-to-surface vertical separation; note that the sample can be at a fixed height and the tip moves vertically with the Z piezo or vice versa. Then, two identical piezo elements provide for displacement along the X- and Y-axes (XY scan stage). Usually, the Z piezo may extend for ~10 μm from its totally compressed position, meaning that features taller or deeper than 5 μm cannot be tracked, while the range of the XY piezos can be up to a few hundreds of microns. A larger scan range is useful to image larger areas, but it also adversely affects lateral resolution and thermal drift. Lateral atomic resolution (to image an atomic lattice) is best obtained with smaller scan stages.

The most used method for monitoring the cantilever movement with the required sensitivity is the optical deflection technique. A small laser diode (typical wavelength in the red or infrared range) is located close to the tip holder, and its beam is focused on the back of the cantilever in such a way that the reflection falls on a four quadrant (ABCD quadrants, see Fig. 2.5) position sensitive photodiode (PSPD). When the tip is at rest, retracted well above the surface, the reflection is positioned exactly at the center of the PSDD; as the cantilever bends upward or downward, the reflection moves up and down on the PSDD, producing different readings in the four quadrants. Also, the torsional motion of the cantilever is monitored and used by the electronics to optimize the feedback loop and to probe other surface features such as surface adhesiveness.

AFM is primarily used to quickly measure the physical dimensions of nanosized objects, such as colloidal or lithography prepared nanoparticles of different shapes, nanowires and nanotubes, two-dimensional flakes, gratings structures, and more complex biological objects. Considering the vast and rapid technological improvements in the industrial fabrication of reliable tips, in the conception of the scanning architecture and detection scheme aiming at more efficient electronic and mechanical noise reduction and artifact suppression, and further considering the constantly evolving and better-performing electronic hardware and controlling computers, AFM has developed into a very user friendly characterization technique with largely automatized measurement routines. Concerning vertical (or z-) resolution, most commercial AFMs can easily perform better than 5 Å and the ultimate value is limited by the piezo element specifications or more realistically by the overall signal stability (i.e., noise level at constant height when the tip is not scanning and when scanning an atomically flat surface); observing atomic steps of crystalline samples or monoatomic layered material is not too demanding, provided adequate installation of the AFM platform in a quiet environment.

Somewhat more delicate is the lateral (xy) resolution: Effectively, this is mostly limited by the size of the tip apex: since the approximately hemispherical apex has finite dimensions (radius of curvature ~10 nm), the profile of any structure appears as a convolution of the actual shape of the structure with the (hemispherical) profile of the tip apex. As a result, the lateral dimensions of an object seen by AFM are always overestimated unless the data are corrected for tip convolution artifacts. If the object dimensions are large compared to the apex, the convolution effect brings only a small error, but if the object has similar dimensions as the tip apex, the error can be very large (Fig. 2.8): For example, 20 nm (diameter) spherical nanoparticles will show up with an almost double diameter; note anyhow that its height (20 nm) will be properly measured and this information can be used to better evaluate the actual lateral dimensions. For more complicated sample shapes, most AFM instrumentation includes routines to evaluate the tip shape and related convolution effects, although these procedures can be more time consuming.

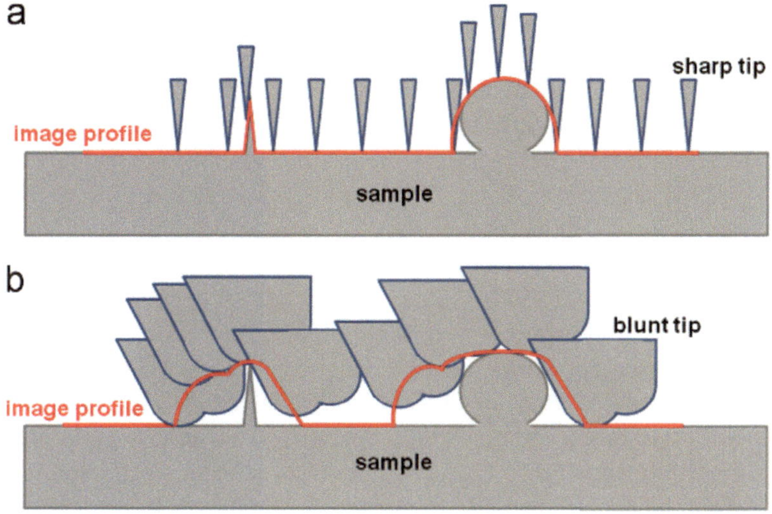

Figure 2.8 Representation of the tip convolution effect. In (a), the left-to-right scanning tip is sharp and with a very small apex; the surface features are well followed, and the AFM trace (profile in red) properly reproduces the surface topography. In (b), the apex has dimensions comparable or bigger than the surface features and the red profile strongly departs from the actual topography; the resulting image is said to have tip artifacts. Note that in both cases, the height of the surface features is correctly estimated. Adapted from Ref. [14].

A straightforward approach to estimate lateral resolution is called the step edge (or knife edge) methodology: A sharp atomic step (~3 Å height difference) imaged by AFM should appear as a Heaviside step function (this would be the case if the tip had no physical dimensions) but rather appears as a sigmoid function. The width of the sigmoid (for practical purpose, this is considered from 0.1 to 0.9 of the normalized height of the step) can be taken as an approximate measure of tip diameter (and twice the best lateral resolution afforded by this tip). In this case, the lateral dimension of the imaged object is negligible with respect to the physical dimensions of the imaging tip. Sharp conical tips with a nominal radius of curvature of only a few nanometers exist when a higher lateral resolution is demanded, but particular care should be taken since these tips are clearly more fragile and easily blunted. Finally, note that the taller the imaged objects, the more severe the tip convolution artifacts: The simple hemispherical tip apex assumption is not sufficient to approximate the tip shape and the resulting convoluted profiles. For objects taller than ~50 nm (or depressions deeper than this value), the actual pyramidal shape of the tip should be taken in account, which can be responsible for surprising alterations of the actual shape of the imaged object.

With respect to SEM or TEM for the observation of nanosized structures, routine AFM cannot be compared with the lateral resolution (and lack of tip artifacts) offered by these techniques. The advantages of AFM are its simplicity of use, rapidity, versatility, cheapness, and minimal need for sample preparation, which make AFM a much-appreciated tool in nanotechnology. While SEM and TEM are most often facilities found at departmental level, AFM platforms are common sights in research laboratories and do not necessitate specifically trained and dedicated technical personnel.

Furthermore, beyond the simple evaluation of the size and shape of nanostructures, several other information on the electrical, magnetic, and mechanical surface properties can be probed at the nanoscale by using specific tips or upgraded instrumentation. The adhesiveness and Young's modulus can be probed at a particular location on the surface by contact AFM force spectroscopy. In short, the tip is cyclically approached and retracted from the surface; beyond the contact point, an increasing force is applied in a controlled fashion by the tip to the surface and then relieved when the tip is retracted. This operation mode has driven tremendous improvements in the AFM electronics, now enabling faster acquisition and treatment of force curves at several points of the surface: In force imaging, the surface topography and the mechanical properties are simultaneously imaged without comprising the overall time required for the scanning. Force spectroscopy and imaging are greatly employed in biology, for example to study modifications in the cell rigidity, which can be related to particular diseases, or to study biomolecular interactions at the single-molecule level by chemically functionalizing the tip with specific antibodies or bioreceptors. In material science, AFM is aptly used for manipulating small molecules or nano-objects and for nano-indentation studies, an example being wear-resistance tests of thin-film coatings on hard materials.

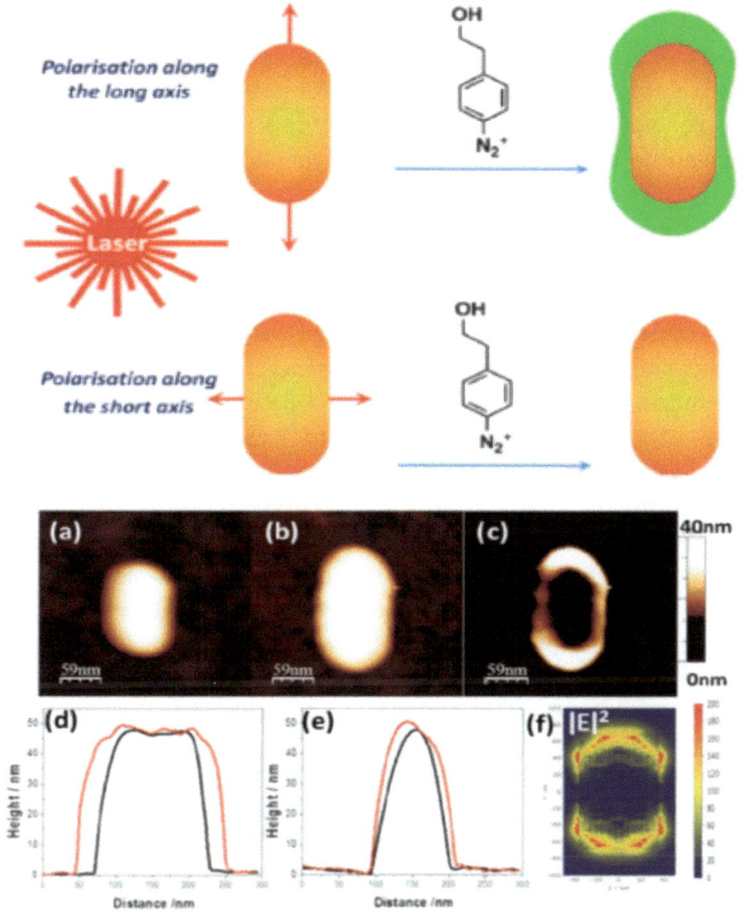

Figure 2.9 (Left) Laser light is used to excite LSP resonances in Au nanorods and to promote the grafting of polymeric aryl diazonium salts at the metal surface. AFM topographic images of the same Au nanorod before (a) and after (b) grafting; image (c) is the subtraction of images (a) and (b). Panels (d) and (e) report the trace profile along the long and short axes of the nanorod, before (black trace) and after (red trace) grafting. An increase in the length of the rod is clearly seen with only a minor change in the width of the nanorod. Adapted from Ref. [15].

The greatest limit of AFM is the lack of chemical sensitivity: There is no information concerning the chemical composition of the sample, although some local differences in the surface composition can be gleaned from the images containing the error signal or, even better, from the adhesiveness and stiffness of the sample when performing force imaging. To relieve this problem, AFM microscopes can be coupled with traditional optical microscopies, such as Raman spectroscopy and fluorescence-based techniques to investigate the sample region close to the tip. An example of the use of AFM in plasmon-promoted chemical reactions is portrayed in Fig. 2.9. The authors used as plasmonic substrate an array of gold nanorods with 165 nm length and 100 nm width fabricated by electron beam lithography, presenting maximum optical extinction at about 785 nm. The nanostructures were incubated in an aqueous solution of hydroxy-ethyl phenyl diazonium salt for 10 s under laser irradiation polarized along the nanorod long axis. A change in the physical dimensions of the individual nanostructures was apparent when comparing the AFM images recorded before and after incubation (Fig. 2.9). The increased length (up to 40 nm) was ascribed to the accelerated and regioselective polymerization taking place at the extremities of nanorods under laser excitation. Due to the anisotropy of the Au nanostructures, strong plasmon excitation takes place only with laser light polarized along the long axis and the corresponding longitudinal resonances mostly localize at the extremities. When repeating the experiment using laser light polarized along the short axis, no change was observed in the sample topography, confirming that the photoinduced deposition of the organic layer is indeed related to the excitation of LSP modes.

References

1. Grand, J., Auguié, B., and Le Ru, E. C. "Combined extinction and absorption UV–visible spectroscopy as a method for revealing shape imperfections of metallic nanoparticles." *Analytical Chemistry* 2019, **91**(22): 14639–14648.
2. Kahraman, M., Mullen, E. R., Korkmaz, A., and Wachsmann-Hogiu, S. "Fundamentals and applications of SERS-based bioanalytical sensing." *Nanophotonics* 2017, **6**(5): 831–852.

3. Haidar, I., Aubard, J., Lévi, G., Lau-Truong, S., Mouton, L., Neuville, D. R., Félidj, N., and Boubekeur-Lecaque, L. "Design of stable plasmonic dimers in solution: Importance of nanorods aging and acidic medium." *The Journal of Physical Chemistry C* 2015, **119**(40): 23149–23158.
4. Haynes, C. L. and van Duyne, R. P. "Nanosphere lithography: A versatile nanofabrication tool for studies of size-dependent nanoparticle optics." *The Journal of Physical Chemistry B* 2001, **105**(24): 5599–5611.
5. Kahl, M., Voges, E., Kostrewa, S., Viets, C., and Hill, W. "Periodically structured metallic substrates for SERS." *Sensors and Actuators B: Chemical* 1998, **51**(1–3): 285–291.
6. Andrews, D., Nann, T., and Lipson, R. H. *Comprehensive Nanoscience and Nanotechnology*. Academic Press, 2019.
7. Quilis, N. G., Lequeux, M., Venugopalan, P., Khan, I., Knoll, W., Boujday, S., de la Chapellebe, M. L., and Dostalek, J. "Tunable laser interference lithography preparation of plasmonic nanoparticle arrays tailored for SERS." *Nanoscale* 2018, **10**(21): 10268–10276.
8. Kociak, M. and Stéphan, O. "Mapping plasmons at the nanometer scale in an electron microscope." *Chemical Society Reviews* 2014, **43**(11): 3865–3883.
9. Bosman, M., Keast, V. J., Watanabe, M., Maaroof, A. I., and Cortie, M. B. "Mapping surface plasmons at the nanometre scale with an electron beam." *Nanotechnology* 2007, **18**(16): 165505.
10. Maver, U., Velnar, T., Gaberšček, M., Planinšek, O., and Finšgar, M. "Recent progressive use of atomic force microscopy in biomedical applications." *TrAC Trends in Analytical Chemistry* 2016, **80**: 96–111.
11. Martin, Y., Williams, C. C., Wickramasinghe, H. K. "Atomic force microscope–force mapping and profiling on a sub 100-Å scale." *J. Appl. Phys.* 1987, **61**: 4723–4729.
12. Seo, Y. and Jhe, W. "Atomic force microscopy and spectroscopy." *Reports on Progress in Physics* 2007, **71**(1): 016101.
13. Alessandrini, A. and Facci, P. "AFM: A versatile tool in biophysics." *Measurement Science and Technology* 2005, **16**(6): R65–R92.
14. Flater, E. E., Zacharakis-Jutz, G. E., Dumba, B. G., White, I. A., and Clifford, C. A. "Towards easy and reliable AFM tip shape determination using blind tip reconstruction." *Ultramicroscopy* 2014, **146**: 130–143.
15. Nguyen, M., Kherbouche, I., Gam-Derouich, S., Ragheb, I., Lau-Truong, S., Lamouri, A., Lévi, G., Aubard, J., Decorse, P., Félidj, N., and Mangeney, C. "Regioselective surface functionalization of lithographically designed gold nanorods by plasmon-mediated reduction of aryl diazonium salts." *Chemical Communications* 2017, **53**: 11364–11367.

Chapter 3

Surface Plasmon Resonance–Mediated Organic Synthesis

Léa Gimeno,[a] Clémence Queffélec,[a] and D. Andrew Knight[b]

[a]*University of Nantes, CNRS, CEISAM, UMR 6230, F-44000 Nantes, France*
[b]*Florida Institute of Technology, 150 West University Blvd, Melbourne, FL 32901, USA*
clemence.queffelec@univ-nantes.fr; aknight@fit.edu

3.1 Introduction

Surface plasmon resonance (SPR) band is an interesting feature of metal nanoparticles (NPs) generated between the electric fields of incident electromagnetic waves and conduction electrons in metal NPs. Based on what has been described in the other chapters of this book and in the literature [1, 2], the localized SPR (LSPR) phenomenon can generate three possible properties for plasmonic NPs: the ability of LSPR to cause local heating, enhance the localized field, and provide charge carriers. These properties are relevant for many photocatalytic reactions. The main goal of this chapter is to

Plasmonic Nanomaterials: Characterization and Applications in Organic Synthesis and Catalysis
Edited by Clémence Queffélec and D. Andrew Knight

ISBN 978-981-5129-09-0 (Hardcover), 978-1-003-47406-7 (eBook)
www.jennystanford.com

describe and understand these processes based on published work in the literature, and we will focus on redox, photothermal, and synthetic reactions.

3.2 Oxidation Reactions

3.2.1 Oxidation of Alcohols, Aldehydes, and Amines

The oxidation of alcohols and aldehydes usually employs strong oxidizing agents such as permanganate or chromates, generating large amounts of toxic waste, and these reactions are often conducted in harmful organic solvents at high temperature. Therefore, efforts have been made toward the development of more benign reagents and catalysts that efficiently oxidize alcohols while still offering high selectivity and the production of less toxic waste.

In that regard, metallic NPs, particularly direct photoexcitation of metal nanoparticles with the appropriate wavelength of light, have received attention in the last decade.

In 2005, in their pioneering work, Tian et al. demonstrated that AuNPs supported on TiO_2 were able to oxidize methanol and ethanol while reducing oxygen under visible light [3]. Since this seminal publication, many studies have been published on oxidation reactions mediated by surface plasmon enhancement.

Tanaka et al. succeeded with the selective oxidation of aromatic alcohols such as benzyl alcohols, substituted with NH_2, NO_2, Cl, etc., to the corresponding carbonyl compounds via the excitation of surface plasmons of AuNPs supported on cerium oxides and using oxygen as the oxidizing agent [4, 5]. Tsukamoto et al. oxidized *sec*-phenethyl alcohol using 5 nm diameter AuNPs located at the interface between the rutile and anatase phases of TiO_2 particles under visible light [6]. Zhai et al. [7] replaced AuNPs with PtNPs deposited on a TiO_2 film, leading to increased reaction rates for a variety of aromatic alcohols.

Plasmon-mediated oxidation of aromatic alcohols has been performed by Hallett-Tapley et al. [8] using simple aqueous colloidal AuNPs with visible light in the presence of H_2O_2 as the oxidizing agent. Rapid and selective oxidation of *sec*-phenethyl to acetophenone was obtained in 20 min when monochromatic 530 nm LEDs provided the light source (Scheme 3.1).

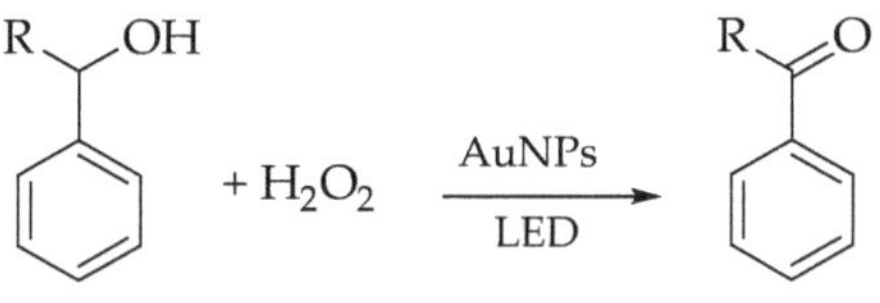

Scheme 3.1 Plasmon-mediated oxidation of benzylic alcohols in the presence of H_2O_2.

Additionally, a laser drop technique was used in which a drop of solution (7–10 μL) was irradiated with a pulsed laser. This resulted in easy observation of the oxidation process by an instant color change solution, but the conversion was lower when compared with LED irradiation.

Regarding the mechanism, they proposed that H_2O_2 is initially transformed into a hydroperoxyl radical HOO● assisted by the excitation of AuNP surface plasmons. Then, the hydroperoxyl radical abstracts one hydrogen from the C–H bond of the alcohol, regenerating H_2O_2 and forming the corresponding cetyl radical, which, by the transfer of an electron to the AuNP, results in a loss of one proton in the final product (Fig. 3.1).

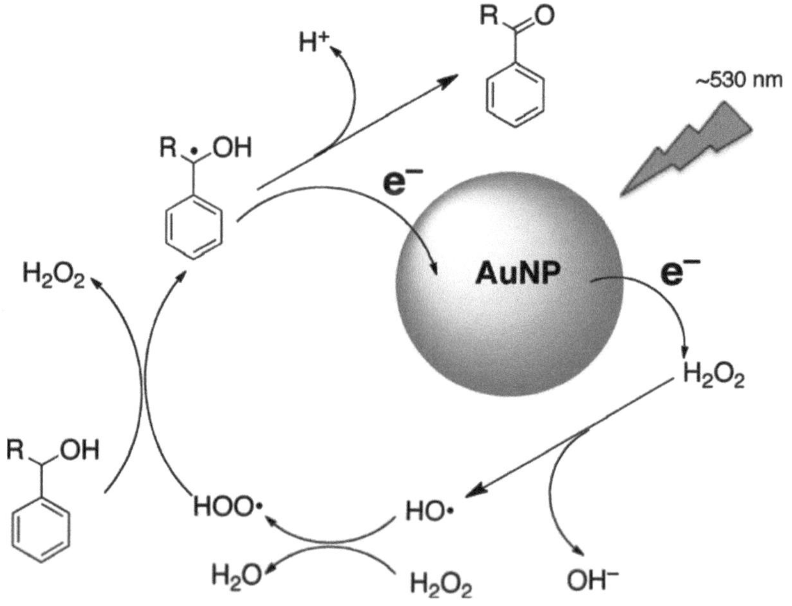

Figure 3.1 Proposed mechanism for the plasmon-mediated oxidation of *sec*-phenethyl alcohol. Adapted from Ref. [8].

Then, AuNPs were supported on hydrotalcite (AuNP@HT), ZnO (AuNP@ZnO), or Al_2O_3 (AuNP@Al_2O_3), and in the absence of a solvent, selectively oxidized benzyl alcohol and *sec*-phenethyl alcohol in the presence of H_2O_2, under 530 nm LED irradiation, in good yields. The reaction was dependent on the type of support employed, with hydrotalcite, the most basic, leading to product formation with 90% yield [9]. In this example, it was proposed that the support might contribute to the mechanism with adsorption of the alcohol substrate and progression of the oxidation reaction being facilitated by the basicity of the support used.

The photocatalytic reactions of 9-anthraldehyde with colloidal or diamond-supported Au nanoparticles (AuNCD) with oxygen or air under green LED were performed by Wee et al. [10], and the best results were obtained with AuNCD (Fig. 3.2). Catalytic reactions were investigated using millisecond real-time resolution using lifetime imaging microscopy (FLIM) and total internal reflection fluorescence microscopy (TIRFM) techniques, and time-resolved events leading to a single nanoparticle were observed, and the subsequent single-molecule conversion of 9-anthraldehyde into its endoperoxide, yielding the final product.

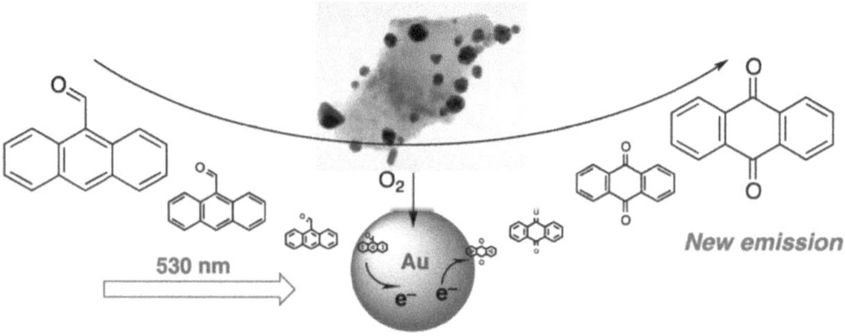

Figure 3.2 Photo-oxidation of 9-anthraldehyde catalyzed by diamond-supported AuNPs with O_2 and under green LED irradiation. Adapted from Ref. [10].

Sarina et al. [11] studied AuPd bimetallic NPs supported on metal oxides. Under irradiation with light, the alloy NPs efficiently enhanced the oxidation of aromatic alcohols under 1 atm of O_2. Similarly, the selective synthesis of imines from the corresponding benzylamines using AuPd bimetallic NP alloys was reported by

Eerdemutu et al. [12]. The reaction yield with an Au:Pd ratio of 1:1.86 in the dark was 36% at 45°C, whereas the yield improved to 95% under a 500 W (400–750 nm) halogen lamp (Fig. 3.3). The mechanism for this reaction using AuNPs and hot carriers has been suggested [13]. The selective oxidative coupling of benzylamine over a UiO-66-NH_2@Au@COF core–shell photocatalyst has also been described. This is the first example of a covalent organic framework supported plasmonic nanomaterial used in organic synthesis [14].

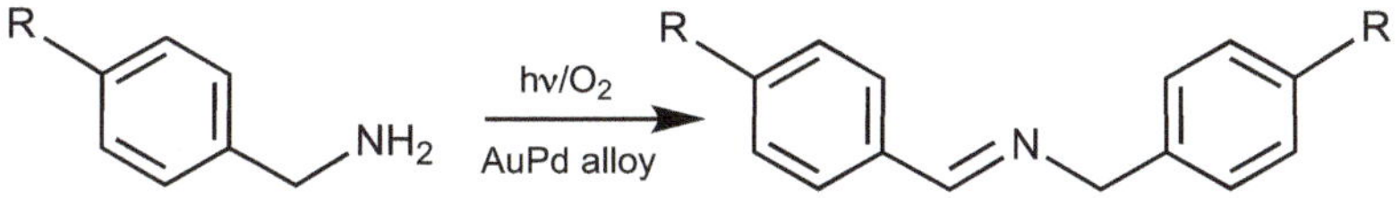

Figure 3.3 Oxidation of amines using gold–palladium alloys under 400–750 nm irradiation. Adapted from Ref. [12].

The dehydrogenation of quinolines using bimetallic nanocatalosomes as plasmonic bilayer shells with interlayer catalytic nanospaces was recently reported by Lee et al. [15]. Other organic transformations, including the Suzuki–Miyaura cross-coupling reaction and alkynyl annulation reactions, were also described. Clean conversions and turnover frequencies greater than at least one magnitude higher than Au nanorod based plasmonic catalysts were noted.

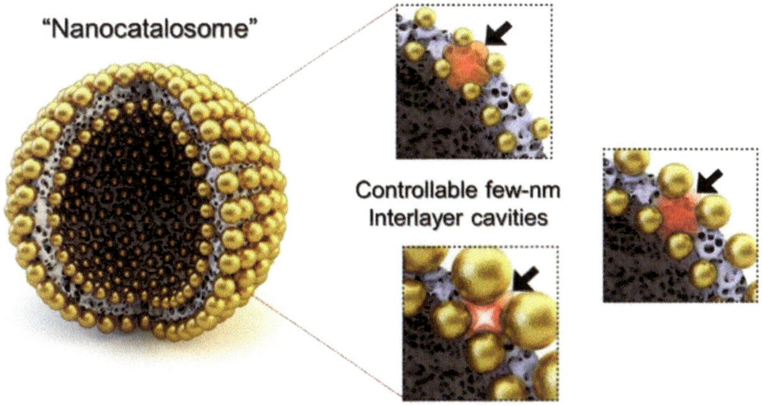

Figure 3.4 Structure of nanocatalosomes used for plasmonic organic transformations. Adapted from Ref. [15].

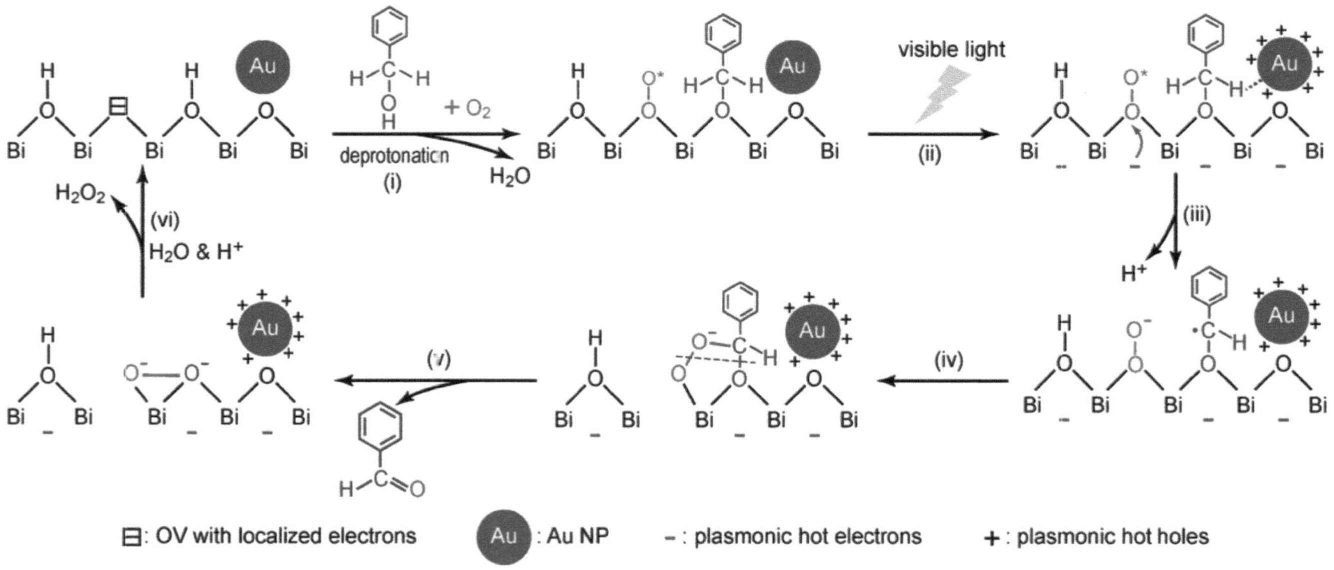

Figure 3.5 Proposed mechanism of selective BO oxidation over Au-BiOCl-OV. Adapted from Ref. [18].

AuPd bimetallic nanowheels [16] or AuPd NPs on TiO_2 nanobelts [17] were also proved to be efficient for the photocatalytic oxidation of benzyl alcohol derivatives. AuNPs (~30 to 35 nm) supported on BiOCl containing oxygen vacancies (BiOCl-OV) also proved to be efficient for the oxidation of benzyl alcohol into benzaldehyde, with O_2 and under visible light [18]. After 8 h, under a 300 W Xe lamp with a 420 nm cutoff filter irradiation, 75% conversion of the aromatic alcohol was observed with a 99% selectivity toward the aldehyde, but under dark condition no conversion was observed. Additionally, with a combination of BiOCl-OV and 30–35 nm AuNPs in the same Au/BiOCl-OV atomic ratio than Au-BiOCl-OV, only 35.1% conversion was observed, and they proposed a mechanism (Fig. 3.5) in which oxygen vacancies on BiOCl likely assist with the trapping and transfer of plasmonic hot electrons to surface-adsorbed O_2, producing $^{\bullet}O_2^-$ radicals (steps I and II), while plasmonic hot holes remaining on the Au surface mildly oxidize benzyl alcohol to the corresponding carbon-centered radicals (step III). Then, a ring addition is proposed to occur between these two radicals on the surface (step IV), which favors the production of benzaldehyde (step V) along with an oxygen atom transfer from O_2 to the product.

For the oxidation of aliphatic alcohols, Au nanorods (AuNRs) supported on TiO_2 have been prepared by Liu et al., and their photocatalytic activities were induced by both transversal and longitudinal plasmon resonance, which promoted the oxidation of isopropanol to acetone [19]. Recently, Nadeem et al. [20] replaced Au with AgNPs supported again on TiO_2, and this plasmonic catalyst allowed the oxidation of ethanol to acetaldehyde. The authors also studied the effects of temperature and the role of LSPR of AgNPs at different temperatures, and it was suggested that at temperatures below 500 K, the SPR of Ag plays a role in the polarization of reaction intermediates involved in the oxidation and/or desorption of the final products. Above this temperature, the SPR from AgNPs is not involved in the reaction as there is enough thermal energy present to promote the reaction. In a systematic study, Kowalska et al. [21, 22] compared the efficiency of AuNPs deposited on various commercial titanium oxides (AuNP@TiO_2) in the oxidation of 2-propanol under irradiation in the visible region of the spectrum. Sugano et al. [23] employed AuCu bimetallic NPs loaded onto TiO_2 as a plasmonic catalyst under visible-light irradiation for the oxidation of isopropanol into acetone. This catalyst performed the oxidation much faster than the individual AuNPs or CuNPs, and the methodology was extended to aromatic alcohols. Furthermore, methanol oxidation to methyl formate was possible with the use of AuPd bimetallic NPs deposited on TiO_2 [24]. The strong enhancement of the catalytic activity was due to the synergism of Au and Pd particles, and the strong interaction between TiO_2 and Pd–Au NPs. Sun et al. described a plasmonic catalyst consisting of Pd–Au NPs supported on a porous carbon matrix. This catalyst exhibited high activity for the oxidation of 2-phenylethan-1-ol and 1-phenylethan-1-ol under light irradiation at 30°C in the absence of a base. The catalyst could be recovered from the reaction and recycled [25].

More recently, Wang et al. [26] studied nanoporous gold (NPG) as a plasmonic catalyst that significantly enhanced electro-oxidation of alcohols. NPG was obtained by de-alloying $Ag_{65}Au_{35}$ (at%) leaves, and the pore/ligament size and surface structure was optimized. The NPG catalyst exhibited high plasmonic catalysis for the oxidation of methanol with a peak current density of 531 $\mu A/cm^2$, which is four times higher when compared to plasmonic catalysis using Au–Ag NPs. An example of a tungsten oxide plasmonic catalyst was reported by Zhu et al. [27]. In this study, plasmonic WO_{3-x} was applied to the alkylation of arenes under red light irradiation with good yields of the coupled product (Fig. 3.6).

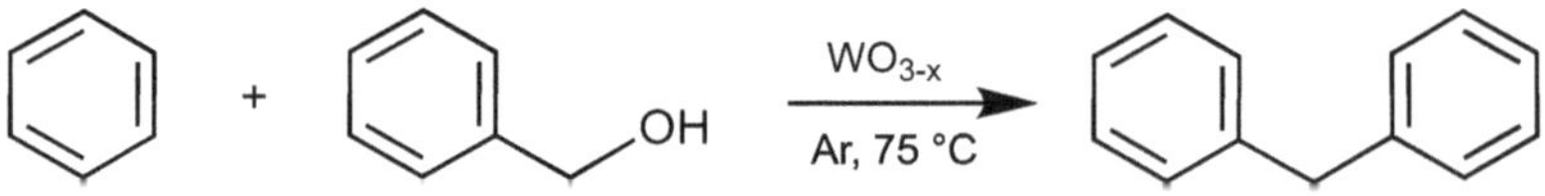

Figure 3.6 Coupling of benzene and benzyl alcohol using a plasmonic tungsten oxide catalyst. Adapted from Ref. [27].

Abad et al. [28] studied the oxidation of allylic alcohols, and the reaction was performed selectively at moderate temperatures using oxygen as oxidant and AuNPs supported on ceria NPs as a catalyst. This catalyst was much more efficient than classical palladium catalysts used in the oxidation of allylic alcohols.

Finally, in 2017, Landry et al. [29] reported the oxidation of aldehydes to carboxylic acids with Ag nanocubes (NCs) at 405 nm and 80°C using 1 atm of H_2 with excellent yield and selectivity.

3.2.2 Oxidation of Alkanes, Alkenes, and Aromatics

In 2011, Christopher et al. [30] used AgNCs supported on α-Al_2O_3 particles for the photocatalytic oxidation of ethylene into ethylene oxide under visible light. These authors also studied propylene epoxidation using CuNPs under the same photochemical conditions. The selectivity was controlled through the reduction of Cu_2O at the surface by varying the light intensity irradiated onto the system [31].

More recently, Zhang et al. [32] studied ethylene epoxidation through the plasmonic photoexcitation of an Ag catalyst at near-ambient conditions.

Surface-enhanced Raman spectroscopy (SERS) was used to monitor the photo-epoxidation of ethylene gas, and the formation of graphene on the surface of AgNPs was observed. The synergistic effect of graphene with AgNPs was proven to be responsible for the epoxidation of ethylene.

Ide et al. studied the selective oxidation of benzene to phenol with AuNPs supported on layered titanate [33]. The rate of this reaction was enhanced when the reaction was performed in aqueous phenol. Subsequently, Zheng et al. [34] studied the same oxidation reaction using noble metals such as Au, Ag, or Pt deposited on TiO_2 particles. Under visible light, Au@TiO_2 exhibited the best selectivity and yield for the oxidation of benzene to phenol when the reaction occurred in aqueous phenol. A proposed mechanism suggests that, following irradiation, an electron transfer occurs from Au nanoparticles (AuNP) to the TiO_2 support. This electron transfer process results in the oxidation of phenoxy anions by the electron-depleted Au, forming phenoxy radicals. These phenoxy radicals, in turn, oxidize benzene to phenol.

Recently, the oxidation of cyclohexane into cyclohexanol was achieved with AgNPs embedded onto hydrated tungsten oxide (WO_3) nanosheets under solar-light irradiation and with the use of tert-butylhydroperoxide as the oxidant [35].

Ag nanoarrays were proven to be efficient for the transformation of alkanes into alkenes under 532 nm irradiation and the oxidization of the methyl group on *para*-methylthiophenol into the corresponding carboxylic acid. This transformation was followed using the SERS technique [36]. When *para*-methylthiophenol was deposited on Ag nanoarrays and irradiated at 532 nm while recording the SERS spectra, two new peaks appeared in the SERS spectra at 1580 and 1690 cm^{-1}. The peak at 1690 cm^{-1} was attributed to C=O and the peak at 1580 cm^{-1} to Ph-COOH/PhCHO, suggesting an oxidation on the plasmonic surface under light illumination. For the transformation of decylamine and 1-butanethiol, a new peak in the SERS spectrum appeared at 3050 cm^{-1}, attributed to the generation of C–H from CH=CH, suggesting that during the SERS measurements, the carbon chains underwent dehydrogenation. A higher laser intensity threshold was necessary for the transformation of alkanes compared to the aromatic one during SERS measurements.

3.2.3 Oxidation of Carbon Monoxide

Lee et al. [37] described the lithographic preparation of a Co_3O_4/Au nanostructure that was efficient for the oxidation of CO with a 25–50% higher catalytic activity under light irradiation compared to the dark. Co_3O_4 is known as a very active catalyst for CO oxidation and, in this case, is even more efficient due to surface plasmon-driven hot carriers created by the absorption of photons on the Au nanostructures (Fig. 3.7a). Figure 3.7b shows the energy band diagram of the Co_3O_4/Au junction and the transfer of hot electrons.

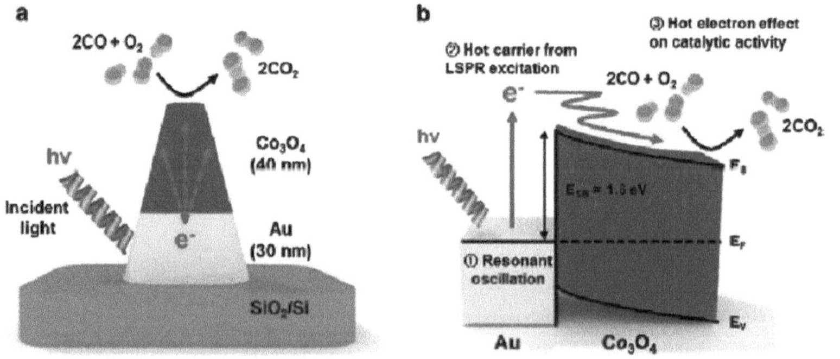

Figure 3.7 (a) Scheme of the patterned Co_3O_4–Au nanostructure and hot electrons; (b) energy band diagram for the Au–Co_3O_4 catalyst. Adapted from Ref. [37].

The same authors [38] developed another catalyst that consisted of hexoctahedral (HOH) Au nanocrystals with Cu_2O clusters. In this report, they studied the structural effect on surface plasmons by preparing Cu_2O at the vertex sites of HOH Au nanocrystals (Cu_2O/Au vertex site) and an HOH Au nanocrystal–Cu_2O core–shell structure (HOH Au@Cu_2O). Both catalysts were found to be active toward CO oxidation under light irradiation, and the Cu_2O/Au vertex site catalyst showed 30–90% higher catalytic activity and the HOH Au@Cu_2O catalyst showed 10–30% higher catalytic activity when compared with the dark conditions. Zhang et al. [39] synthesized AuNPs supported on ZIF-8 (a metal–organic framework) modified TiO_2 (Au/ZIF-8-TiO_2). In this catalyst, ZIF-8 was distributed on the surface of TiO_2 microspheres, enhancing the surface roughness of the latter, thus enabling the dispersion of AuNPs onto the support due to the large surface area and porosity. This nanocatalyst was more efficient in CO oxidation under visible-light irradiation than Au/TiO_2, and it was suggested that ZIF-8 acted as (a) an electron donor to enrich the surface electron densities of both Au and TiO_2, resulting in a better adsorption and activation of CO and O_2, thus enhancing the oxidation reaction, and (b) an electron-transfer mediator that promoted the electron transfer between Au and TiO_2 induced by the LSPR of AuNPs under irradiation.

3.3 Reduction Reactions

The reduction of nitro compounds into amino or azo compounds using plasmonic material under irradiation has been extensively reported, and in the following section, only a few selected examples will be mentioned here.

3.3.1 Reduction of Nitro-Containing Compounds to Azo-Containing Compounds

The reduction of nitro-group molecules with the assistance of hot electrons from noble metal surface plasmon decay has been largely studied for the formation of azo compounds [40–43].

Direct reduction of nitroaromatics to azobenzenes was accomplished using the SPR of AuNPs supported on zirconium oxide (AuNP@ZrO_2) illuminated with visible light or ultraviolet light at 40°C [44]. Ag–Cu alloy NPs supported on ZrO_2 were used for the selective hydrogenation of aromatic nitro compounds to their corresponding azoxy compounds (Scheme 3.2) [45].

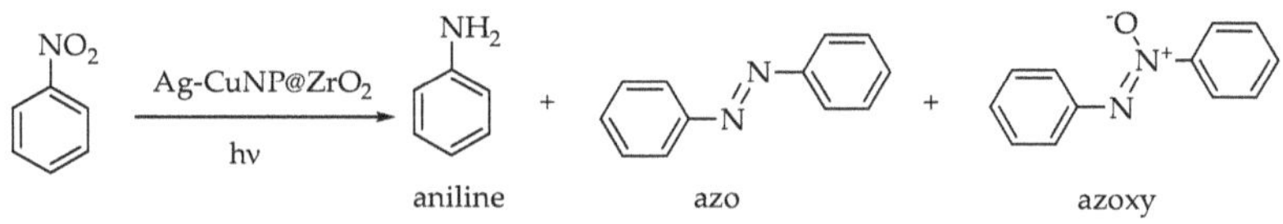

Scheme 3.2 Reduction of nitrobenzene into azoxy compounds.

The Ag–Cu molar ratio of 4:1 exhibited the best performance. The authors did demonstrate that tuning the wavelength influenced the product selectivity from aniline to azoxybenzene while Cu was found to be responsible for the product selectivity shift from azobenzene to azoxybenzene. The reduction of nitrobenzene to azobenzene and hydroazobenzene with high yield and selectivity using an Au/TiO_2 catalyst was described by Liu et al. [46]. In this paper, the reaction mechanism was investigated using electron probe resonance (EPR) and an Au–H intermediate species was suggested as being responsible for the reduction process (Fig. 3.8). The authors intriguingly suggest the potential application of plasmonic catalysis on an industrial scale.

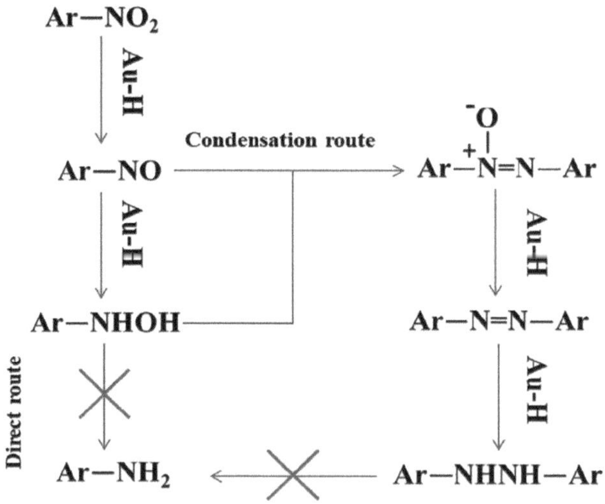

Figure 3.8 Proposed mechanism for reduction of nitroarenes involving gold-hydride species. Adapted from Ref. [46].

Ding et al. [47] studied the coupling reaction of p-nitroaniline, on silver nanostructures and under ambient pressure using SERS.

These authors found that 4,4′-diaminoazobenzene was formed via the selective reduction of the nitro group on *p*-nitroaniline (Scheme 3.3). Additionally, it was shown that graphene–metal hybrid nanostructures also performed well for this selective reduction.

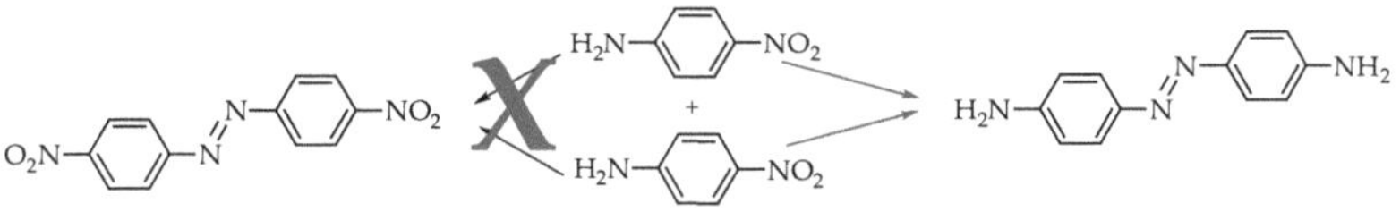

Scheme 3.3 Illustration of the coupling of p-nitroaniline into 4,4′-diaminoazobenzene on AgNPs.

Ke and collaborators [48] showed that AuNPs supported on a metal oxide when irradiated in the visible region of the spectrum were efficient photocatalysts. The highest performance support was cerium oxide, and AuNP@CeO_2 exhibited the highest catalytic activity in the reduction of nitroaromatics to azo compounds, hydrogenation of azobenzene to hydrazobenzene, reduction of ketones to alcohols, and deoxygenation of epoxides to alkenes at room temperature, employing isopropanol as the reducing agent. When irradiated with light, AuNP@CeO_2 abstracts a hydrogen atom of isopropanol, forming AuNP–H on the surface, which is the active species for the reduction process of the functional groups N=O, N=N, C=O, double bonds, and epoxide bonds.

Kang et al. [49] monitored the plasmon-driven chemical reaction on AgNPs of *para*-nitrothiophenol (pNTP), which undergoes dimerization to form *para*-dimercaptoazobenzene (DMAB) by SERS. The excitation wavelength and the laser power had substantial effects on the speed of the reaction (Fig. 3.9). The same catalytic reaction using Au/Ag core–shell nanostars was reported by Kaur et al. [50].

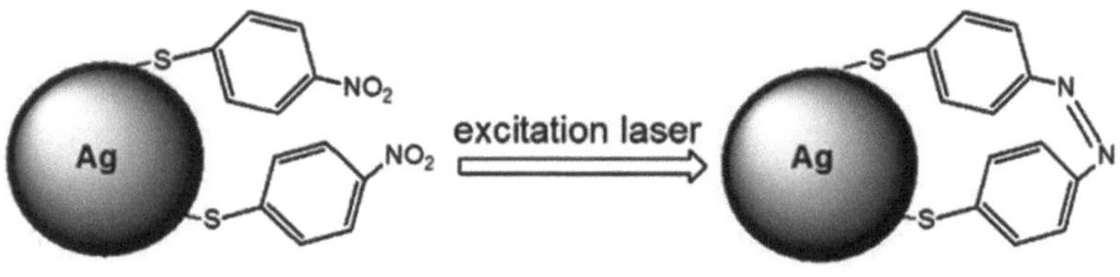

Figure 3.9 Illustration of the dimerization of *p*-nitrothiophenol into dimercaptoazobenzene on a single Ag particle. Adapted from Ref. [49].

Finally, Koopman et al. [51] studied the dimerization of 4-nitrothiophenol (4NTP) to 4,4′-dimercaptoazobenzene under light irradiation with Au nanoflowers. They established that the transfer of hot electrons lowers the initial activation barrier, but the reaction is still then limited by the thermally activated dimerization step. This suggests that plasmonic NPs can play a role in the various steps of a reaction employing different mechanisms simultaneously.

It is interesting to mention here that the transformation of nitro compounds has commonly been observed by SERS and SERS-induced reactions. Tip-enhanced Raman scattering (TERS) has also been utilized to induce organic transformations [42, 52–56], and this will be discussed in Chapter 5.

3.3.2 Reduction of Nitro-Containing Compounds to Amines

One of the first reports on the use of AgNPs for the reduction of nitro aromatic compounds to the corresponding amines was published by Pradhan et al. [57]. Since then, many reports have described this reduction reaction with various nanocatalysts. A number of silver [8, 24–26] or silver alloy [58, 59], gold [60–64] or gold alloy [65–69] nanocatalysts using H_2 or sodium borohydride were reported to possess photocatalytic activities for the reduction of nitrophenol compounds under visible light.

For example, reduction of *para*-nitrophenol by sodium borohydride, catalyzed by gold nanostructures, led to the formation of *para*-aminophenol [62]. Au-based nanocages were found to be more active than nanoboxes or NPs. This reduction reaction has been investigated as well on Au–Ag alloy NPs by in situ real-time quantitative SERS, which allows the signal detection of both the intermediate and final products [59]. The selective hydrogenation of 4-nitrostyrene to the corresponding aniline using ammonia:borane and a porous triple-shell AuNP modified with sulfonated polystyrene provided excellent selectivity of the unmodified counterpart. The authors suggest that the number of shells plays an important part in the enhancement of catalytic activity [70].

Gold/copper alloy NPs [66] improved the selectivity of the reduction toward nitro groups under visible light, avoiding byproducts formation, e.g., compounds containing the azo, azoxy, or hydrazo groups. More recently, Verma et al. [71] studied the reduction of nitro aromatic compounds using glycerol as a green reductant and AgNPs functionalized with curcumin, which acted as a stabilizing agent. They observed that under visible light, an increase in the amount of curcumin on the AgNP surfaces was beneficial for the catalytic activity suggesting that curcumin might actively participate in the photocatalytic reaction. Thus, they proposed a mechanism shown in Fig. 3.10. The enhancement of the photocatalytic reaction was about five times that of the simple thermal reaction.

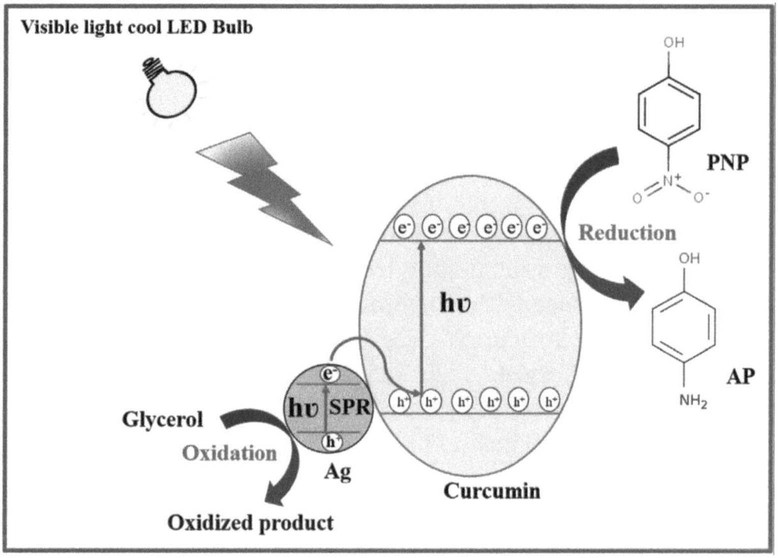

Figure 3.10 Proposed mechanism for the reduction of *p*-nitrophenol with AgNPs functionalized with curcumin and under visible light irradiation. Adapted from Ref. [71].

Barbosa et al. [72] studied the reduction of 4-nitrophenol using either $H_{2(g)}$ or $BH_4^-{}_{(aq)}$ as reducing agents and AuNPs deposited on TiO_2 or SiO_2 colloidal spheres under LED irradiation. An increase in activity was observed when Au/SiO_2 NPs were used. When Au/TiO_2 was employed as a catalyst, plasmonic enhancements were demonstrated when $H_{2(g)}$ was added as the reducing agent, but when $BH_4^-{}_{(aq)}$ was used as the reducing agents, negative catalytic effects were observed due to charge transfer of plasmon-excited hot electrons from Au to TiO_2. Indeed, under dark conditions, after

1 h, the conversion of the reaction reached 100% for the Au/TiO_2 catalyst, while only 35% conversion was obtained for Au/SiO_2 NPs. Under LED irradiation, the conversion reached 78% for the Au/SiO_2 catalyst, while for the Au/TiO_2 catalyst, the conversion dropped to 73%. Thus, it seems that the nature of the reducing agent and the occurrence of charge-transfer processes at the interface can be responsible for different plasmonic catalytic activities.

RuPd alloy NPs supported on an hydrogen-doped molybdenum oxide support (H_xMoO_{3-y}) proved to be effective in the reduction of *p*-nitrophenol under visible-light irradiation [73]. The oxide support has a plasmonic band localized at 639 nm and acts as an electron donation center, while the RuPd alloy NPs act as the active species. Typically, the reduction of nitroaromatics involves the utilization of hydrogen or a borohydride reducing agent. However, an alternative approach can be found employing formate as a hydrogen donor [74]. The authors describe the use of anisotropic Pd–Au nanorods for the plasmon-enhanced selective hydrogenation of nitrobenzene to aniline under visible-light irradiation. The plasmon-induced interaction between the hydrogen donor, nitrobenzene, and plasmonic nanostructures was probed using in situ single-particle luminescence. Light-induced adsorption enhancement of nitrobenzene was shown to be responsible for the plasmon-enhanced catalysis (Fig. 3.11).

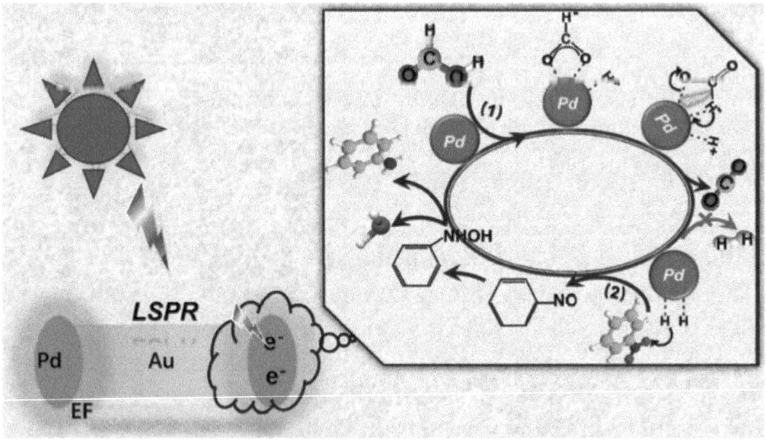

Figure 3.11 Light-induced adsorption enhancement of nitrobenzene is responsible for plasmon-enhanced catalysis. Adapted from Ref. [74].

3.3.3 Reduction of Aldehydes and Ketones

Hao et al. [75] developed an Au/SiC catalyst that was shown to be very active in the selective hydrogenation of α,β-unsaturated aldehydes to the corresponding unsaturated alcohols with high TOF and selectivity under visible-light irradiation. 2-propanol was used for the transfer hydrogenation, and cinnamaldehyde was hydrogenated to cinnamyl alcohol, while 2-propanol was transformed into acetone. The selectivity was 100% for the production of alcohol under visible-light irradiation at 20°C (Fig. 3.12).

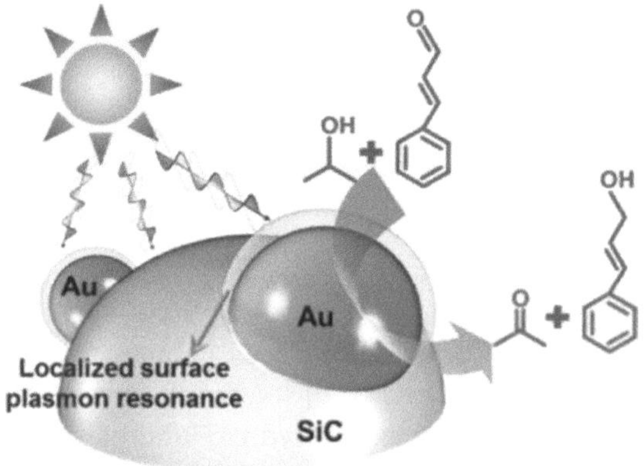

Figure 3.12 Selective hydrogenation of cinnamaldehyde to cinnamyl alcohol with 2-propanol using SiC-supported Au. Adapted from Ref. [75].

When the SiC support or AuNPs were tested alone under irradiation, cinnamyl alcohol was obtained with low yields and selectivity. Finally, other α,β-unsaturated aldehydes were reduced to the corresponding unsaturated alcohols with high conversion and selectivity.

Landry et al. [29] studied the hydrogenation of carbonyl compounds with silver nanocubes under visible-light irradiation. The reduction was performed at 405 nm and 80°C using 1 atm of H_2 with excellent yield and selectivity toward C=O over C=C bonds.

3.3.4 Miscellaneous Reduction Reactions

Resazurin is an interesting dye because its reduction can be followed by either colorimetric or fluorometric methods. In the colorimetric method, the color of resazurin changes from blue (λ_{max} = 603 nm) to pink (λ_{max} = 570 nm). Regarding its fluorescence, the change is even more visible since resazurin is very weakly fluorescent compared to the strong emission of the resorufin fluorescence (λ_{em} = 582 nm) (Fig. 3.13).

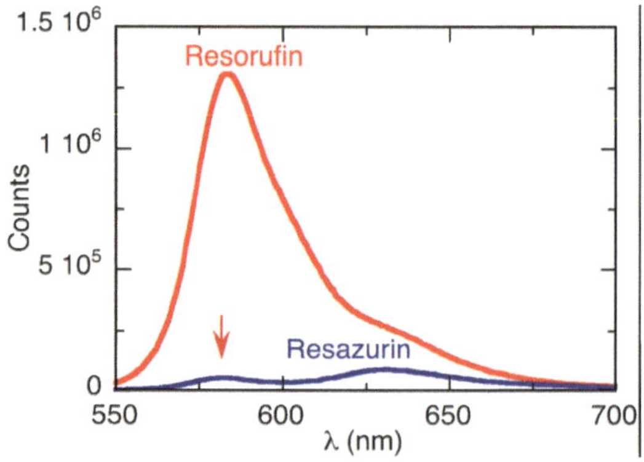

Figure 3.13 Fluorescence emission spectra for resazurin (blue) and resorufin (red). Adapted from Ref. [76].

In 2011, Bueno-Alejo et al. [76] studied the reduction of resazurin to resorufin in the presence of hydroxylamine and AuNPs under plasmonic excitation. In this example, a laser drop technique was used, which consists of irradiating a drop of solution (7–10 μL) with a pulsed laser, and this technique resulted in the straightforward observation of the reduction process via an instant color change in the solution, as seen in Fig. 3.14.

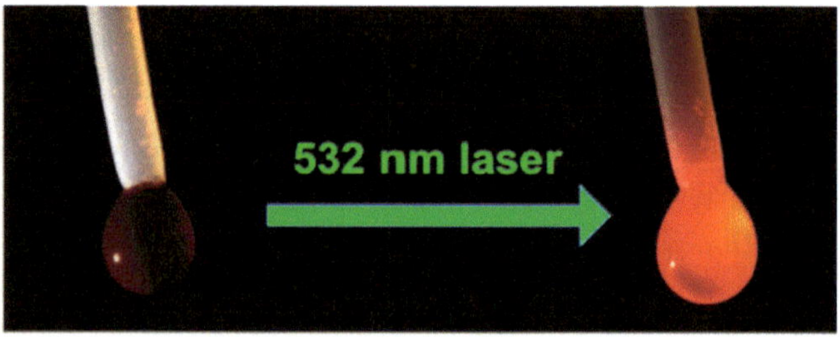

Figure 3.14 Photos of a drop (~10 μL) containing resazurin, AuNP, and NH_2OH before (left) and during (right) excitation using the laser-drop technique (λ_{exc} = 532 nm). The change in the drop color is a consequence of the formation of resorufin. Adapted from Ref. [76].

This catalytic photoreduction proceeded with high efficiency when a pulsed laser was used (λ = 532 nm; 8 ns). When the laser is replaced with light-emitting diodes, the reaction proceeds more slowly (Scheme 3.4). The catalytic reaction is thermally activated by AuNPs via the high temperatures reached by plasmonic excitation.

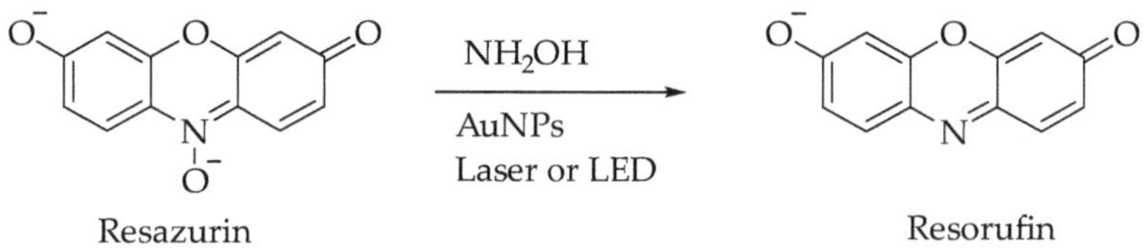

Scheme 3.4 Resazurin reduction to resorufin in the presence of hydroxylamine and AuNPs.

The authors proposed a mechanism for this reaction in which gold nanoparticles behave as both a nanoheater, through the excitation of the resonance band plasmonic, and a catalyst. In a series of recent works, Chen et al. [77–79] clearly showed that gold nanoparticles can act as effective catalysts in this reduction. In addition, dependence on fluorescence with the hydroxylamine concentration indicates that this reductant must be involved in the rate-determining step. Thus, hydroxylamine either reaches the resazurin molecule at the reactive site on the AuNP during the brief period of plasmon-induced heating or assists in the departure of the reduced product at the AuNP surface.

Phenylacetylene was reduced under visible light with H_2 as the reducing agent and Au@Ag@Pt core–shell nanocatalyst or an Au@AgPt nanorattle architectures [80]. Under light irradiation, after 8 h, only a slight increase in the phenylacetylene conversion was observed for the core–shell Au@Ag@Pt/SiO_2 NPs, while a significant increase was observed for the Au@AgPt/SiO_2 nanorattles (from 44 to 98%). The authors then focused on the selectivity of the reaction because the reduction of phenylacetylene can lead to styrene or ethylbenzene compounds. Under irradiation, Au@Ag@Pt/SiO_2 core–shell and Au@AgPt/SiO_2 nanorattles were found to be more selective toward styrene with selectivity that corresponded to 87% and 84%, respectively. The nanorattle catalyst is of great interest because it leads to better selectivity and reactivity, and this might be explained by the fact that it presents a higher active surface and can concentrate high electric field intensities at the surface. It also enables efficient energy transfer to the non-plasmonic metal (Pt), where the LSPR-excited hot carriers lead to the enhancement of activity and selectivity under irradiation.

Conversion of renewable biomass resources is highly desirable especially with non-noble metal catalyst. Wang et al. [81] developed an original catalyst based on copper, a non-noble metal for the study of the hydrogenation reaction. Copper NPs were encapsulated in a few layers of carbon, and this catalyst exhibited high catalytic performance for the hydrogenation of furfural into furfuryl alcohol under visible light. In addition, the heterogeneous catalyst was stable and was reused for several catalytic runs without a significant loss in activity. Finally, the reduction of carbon dioxide to methane using a bimetallic Ag–Ni catalyst under 405 nm laser light was studied, and photocatalytic activity was found to be dependent on the extent of bimetallic interactions [82].

3.4 Photothermal Reactions

Decomposition of dicumyl peroxide can be achieved thermally, but high temperatures are required (>140°C) due to the high activation energy (34.3 kcal/mol), which is necessary for breaking the peroxide bond. The reaction may also be promoted photochemically, but the reaction requires ultraviolet irradiation. In both cases, the cumyloxy radicals formed after the homolytic breaking of the O–O bond result in the formation of acetophenone or 2-phenyl-2-propanol (Scheme 3.5).

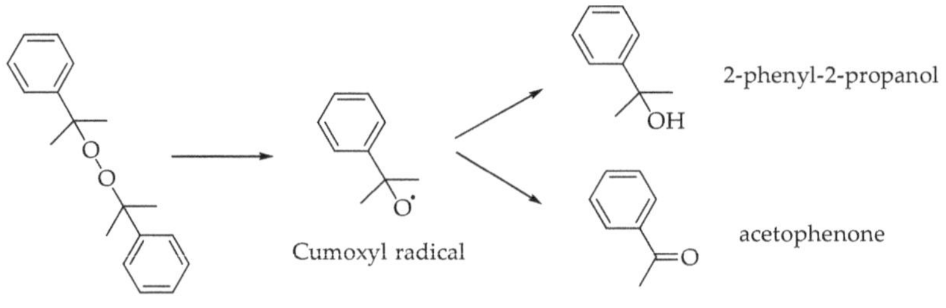

Scheme 3.5 Cumiloxyl radical reactions.

Fasciani et al. [83] studied this decomposition using AuNPs and the laser drop technique (mentioned previously) with a pulsed laser excitation at λ = 532 nm. AuNP irradiation at 532 nm in aqueous solution containing dicumyl peroxide and methanol resulted in the complete decomposition of peroxide in less than 1 min with the formation of 2-phenyl-2-propanol. The authors suggested that photothermal effects led to the exposure of peroxide molecule to temperatures close to 500°C for submicrosecond times. This example highlights the possibility of performing reactions that require high energies at room temperature.

Bakhtiari et al. [84] grafted a fluorescent dye onto AuNPs and studied the photothermal release. The retro Diels–Alder reaction was the focus of the study since the temperature at which it is activated can be easily tuned by modifying the structure of the diene and dienophile. Thus, 7-oxa-bicyclo-[2.2.1]hept-5-ene-2,3-dicarboxylic imide containing fluorescein was anchored onto SiO_2@ AuNPs (core = silica and the sell = Au). Of particular interest is that the fluorescence emission from the dye is quenched by the gold layer when close to the surface and as soon as the dye is released from the nanoparticle surface and free in the medium, the dye is emitted easily (Fig. 3.15).

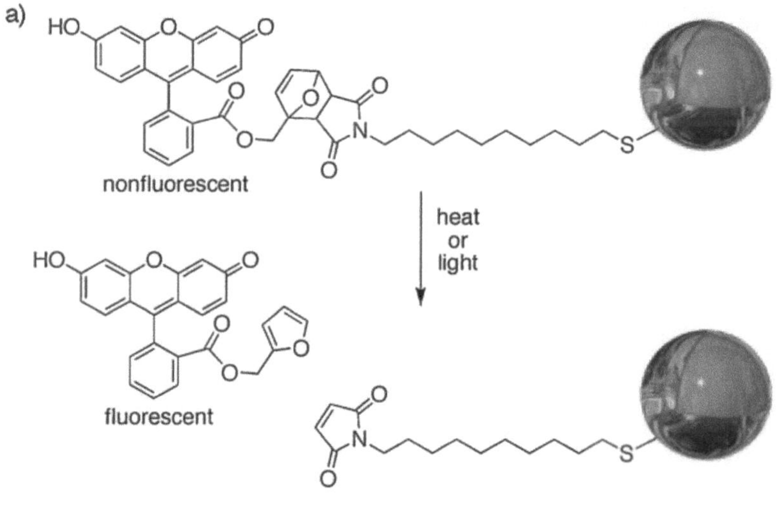

Figure 3.15 Release of fluorescein dye from the surface of a core–shell NP (silica core/Au layer) using photothermal effects. Reproduced from Ref. [84].

The authors focused on core–shell nanoparticles of ca. 200 nm diameter that possess two SPR bands (700 and 1000 nm) in the NIR region and studied the retro Diels–Alder using a pulsed laser at 800 nm. The rate of photothermally induced release from the core–shell nanoparticle was around four times higher than the thermally induced release. Finally, the photothermal synthesis of the synthetically important α-cyanoacrylonitriles was reported by Liang et al. [85]. A heterogeneous catalyst composed of AuNPs supported onto defect-rich h-boron nitride sheets promoted the photothermal (Xe lamp, 18 h) reaction of benzaldehyde and malonitrile to give the corresponding α-cyanoacrylonitrile in high yield compared to the dark reaction (Fig. 3.16).

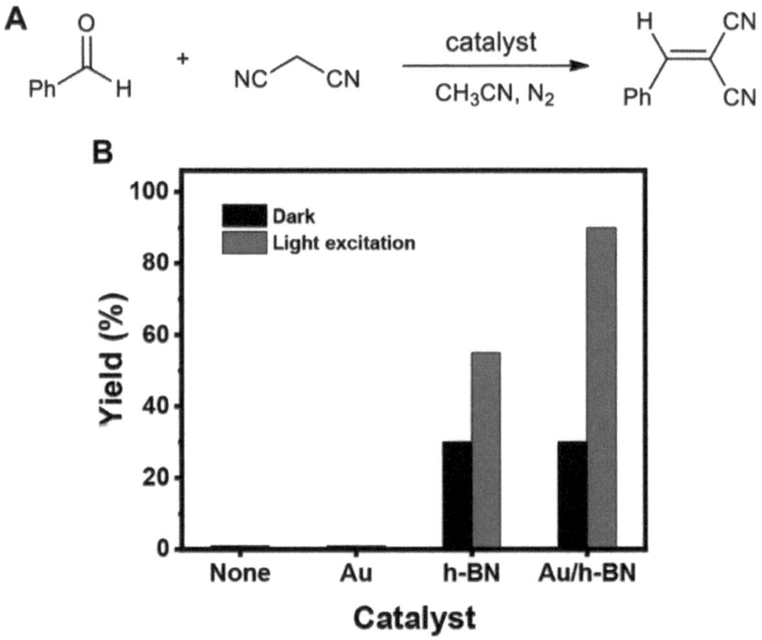

Figure 3.16 Photothermal reaction of benzaldehyde and malononitrile under light and dark conditions. Adapted from Ref. [85].

3.5 Miscellaneous Chemical Reactions

3.5.1 Suzuki–Miyaura Reaction

The Suzuki–Miyaura reaction was discovered by A. Suzuki in 1979 and the Nobel Prize for this discovery was shared in 2001 with R. Heck and E. Negishi. The reaction is a cross-coupling between an aryl boronic acid and an aryl halide leading to the formation of biaryl compounds, catalyzed by a palladium(0) catalyst.

Sarina et al. replaced the classical catalyst with a gold–palladium NP alloy supported on zirconium oxide [11]. Under light irradiation, the alloy NPs strongly absorb light and efficiently enhance the conversion of the Suzuki–Miyaura cross-coupling. The intensity and wavelength of incident light have a strong impact on the catalytic activity. For example, the irradiation increased the product yield of the Suzuki coupling at 30°C to 96% (from 37% in the dark) with an optimal Au/Pd molar ratio of 1:1.86 (Scheme 3.6).

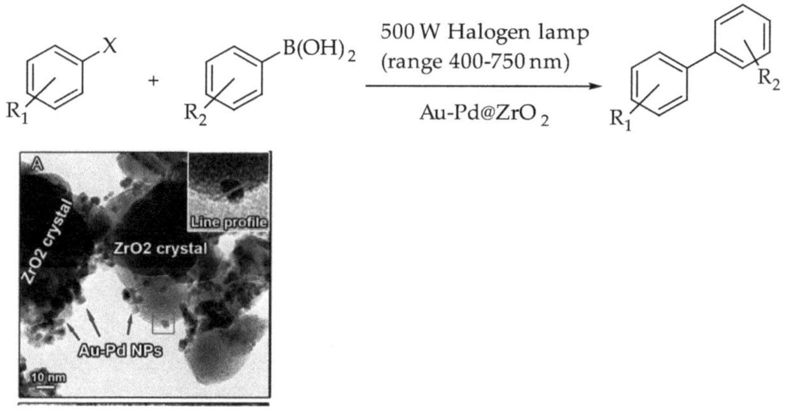

Scheme 3.6 The Suzuki–Miyaura reaction. Adapted from Ref. [11].

Han et al. [86] used a gold–palladium NP alloy supported on titanium oxide, and under blue LED irradiation, the yield of the reaction reached 98% after 5 h versus 17% in dark conditions with 1 mol% of nanocatalyst. Moreover, if the quantity of nanocatalyst is decreased to 0.05 mol%, the yield remains constant at ~96%. The use of such a low amount of catalyst is very important in terms of economic and ecologic aspects.

Nanorods possess large extinction cross sections and have tunable longitudinal plasmon wavelengths. In addition, PdNPs are excellent catalysts for carbon–carbon bond formation. Wang et al. [87] decorated Au nanorods with PdNPs, and under the illumination of an 809 nm laser at 1.68 W, the Suzuki reaction was induced and accelerated through both plasmonic photocatalysis and photothermal conversion. The yield of the Suzuki coupling reaction was ~2 times higher compared to the reaction that was heated via conventional methods to the same temperature. For example, the best efficiency was observed at 62°C in which 99% photothermal conversion was obtained, while 55% yield was obtained under isothermal heating (Scheme 3.7).

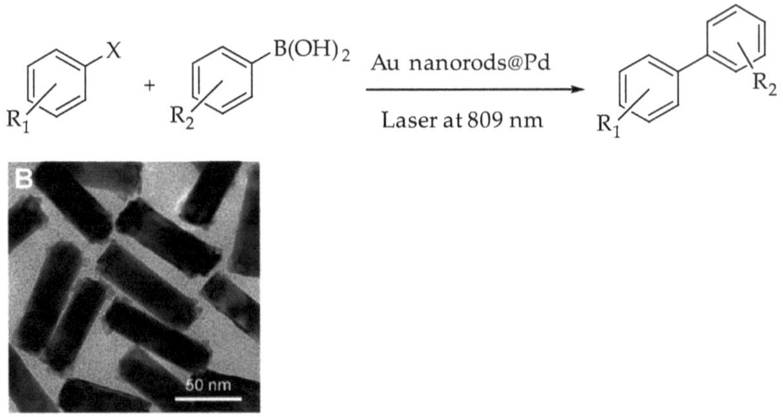

Scheme 3.7 The Suzuki–Miyaura reaction. Adapted from Ref. [87].

Au nanorods were also used in the Suzuki-Miyaura coupling by Yoshii et al. [88]. The authors deposited Pd clusters onto homogeneous reduced graphene oxide (rGO) layer-coated Au nanorods, and the resulting catalyst exhibited 3.8 times higher activity under visible-light irradiation (filter at 420 nm) compared with the dark reaction. The conversion was 56% at 30 min (Scheme 3.8). The authors noticed the importance of the rGO layer because without this layer between the palladium clusters and the Au nanorods, the activity of the Pd–Au catalyst was only 1.5 times higher than in the dark. These results suggest that the rGO layer provides high performance for the SPR-enhanced reaction. It is important to point out that even though enhancement was observed under light irradiation, yields of the reaction obtained were low, in the range of 8–12%.

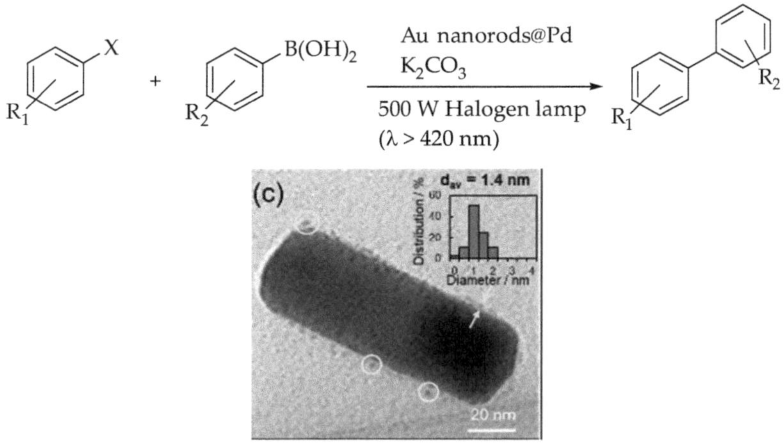

Scheme 3.8 The Suzuki–Miyaura reaction. Adapted from Ref. [88].

Zhu et al. [89] studied the influence of a plasmonic anisotropic nanocatalyst that consisted of a gold nanopyramid core covered at the extremities with silica and on the sides with palladium. This nanocatalyst was employed in the Suzuki–Miyaura cross-coupling, and the reaction was conducted in water at 50°C. After irradiation with an 808 nm laser for 1 h, the conversion of the reaction was around 85% versus 65% under dark conditions (Scheme 3.9).

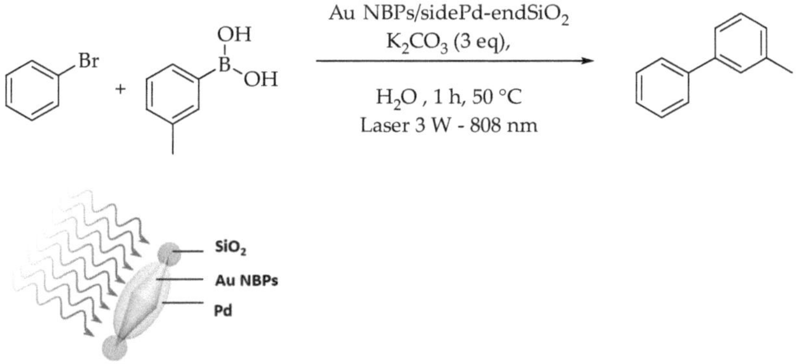

Scheme 3.9 The Suzuki–Miyaura reaction with Au nanopyramids coated with silica and palladium. Adapted from Ref. [89].

Recently, Pd nanoflowers were synthesized via the reduction of Pd(II)(acac)$_2$ in ethanol under a CO atmosphere [90]. These nanoflowers displayed a broad absorption plasmon band in the visible and near-IR ranges, and their plasmonic properties under visible-light irradiation in Suzuki–Miyaura cross-coupling reactions were studied. The reaction was carried out using iodobenzene, phenylboronic acid, and Cs_2CO_3 as the base under visible-light irradiation, and the results were compared with the reaction conducted in the dark. Under light irradiation (300 W Xe lamp equipped with a water cell filter to absorb the near-IR radiation, and a 475 nm cutoff filter to avoid direct photoactivation of the aromatic compounds), the conversion was 96% versus 53% in the dark after 240 min. The authors explained that the plasmonic effect increases the reactivity of the palladium nanostructures, which induces heat and hot electrons on the Pd surface. Koohgard et al. [91] studied PdNPs deposited onto TiO_2 through a photodeposition method under solar light that exhibited plasmonic properties. In aqueous media (water:PEG = 1:1), 4-methoxyiodobenzene reacted with phenylboronic acid to give the final product with 91% yield under visible-light irradiation (400–750 nm, LED, 15 W, 28°C) compared to 22% yield in dark conditions after 4 h. Photogenerated electron–hole pairs formed under visible-light irradiation are the main parameters of activation of the different substrates in the Suzuki–Miyaura reaction. Gangishetty et al. [92] used AuPd bimetallic nanotriangles, and under green LED irradiation, the Suzuki–Miyaura reaction rate was improved compared to the dark reaction after only a few hours, and this enhancement was mainly due to plasmonic heating effects, with a minor contribution from plasmonic hot electrons. Nemygina et al. [93] synthetized a nanocatalyst composed from 20 nm core AuNPs surrounded by a shell of 1 nm palladium NPs and immobilized within hyper cross-linked polystyrene. Under visible-light irradiation using a 300 W filament lamp, the product yield of the Suzuki–Miyaura coupling was 1.3 times higher than in the dark (from 55.3% to 71.6%). Furthermore, Rohani et al. [94] designed a hydrogenated urchin-like yolk@shell TiO_2 structure decorated with Au and Pd nanoparticles (HUY@S-TOH@AuPd), which was used in the Suzuki–Miyaura reaction. Under 300 W xenon lamp irradiation, 4-methyliodobenzene reacted with phenylboronic acid in the presence of this nanocatalytic system with a 94% yield after 1 h, while under dark conditions, no product was formed. Very recently, Li et al. [95] focused on TERS to study the Suzuki–Miyaura reaction. In that regard, a monolayer of one reactant was deposited on the surface of Au@Pd NPs and then the second reactant was deposited on the scanning probe. Then the probe is brought close to the sample surface, which is illuminated continuously by a laser, and the reaction takes place and is monitored by Raman spectroscopy. Thus, they could probe within seconds the reactivity of different aryl halides under irradiation. This TERS technique allows fast, robust, and reliable analysis and understanding of molecular reactivity for plasmon-enhanced reactions. Finally, AuPd bimetallic nanowheels used in the oxidation of benzyl alcohol were proven to be efficient catalysts in the Suzuki coupling reaction [16]. Li also reported the encapsulation of core–shell Cu@Ni NPs into the zeolitic imidazolate framework ZIF-8 for the photocatalytic Suzuki coupling reaction [96]. This is one of the few examples of the incorporation of nanoparticles into metal–organic frameworks and allows the combination of high porosity, shape-selectivity, and catalytic enhancement.

It is important to note that the C–B bond cleavage deboronation is a side reaction that can lower the Suzuki coupling reaction yield. Very recently, a protodeboronation reaction [97] was achieved using AuNPs and planar gold substrates under 785 nm light irradiation. The model reaction studied was the protodeboronation reaction from 4-mercaptophenylboronic acid (MPBA) to benzenethiol (BT) (Scheme 3.10), and this reaction occurs upon plasmon excitation in nanogaps between the Au substrate where a self-assembled monolayer (SAM) of MPBA was deposited followed by AuNP deposition on the SAM. A schematic representation of the setup is presented in Fig. 3.17.

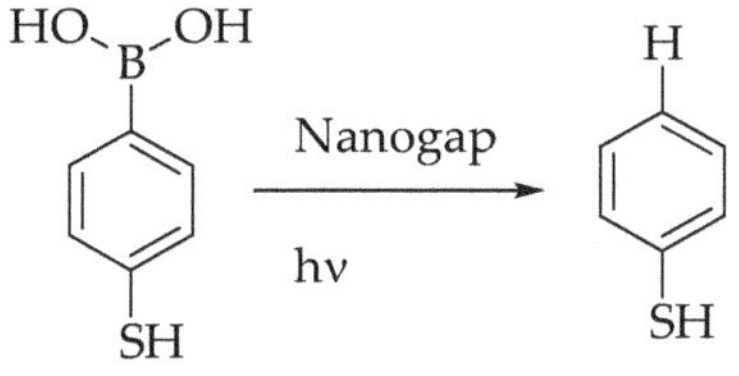

Scheme 3.10 Protodeboronation of MPBA to BT in nanogaps upon plasmon excitation.

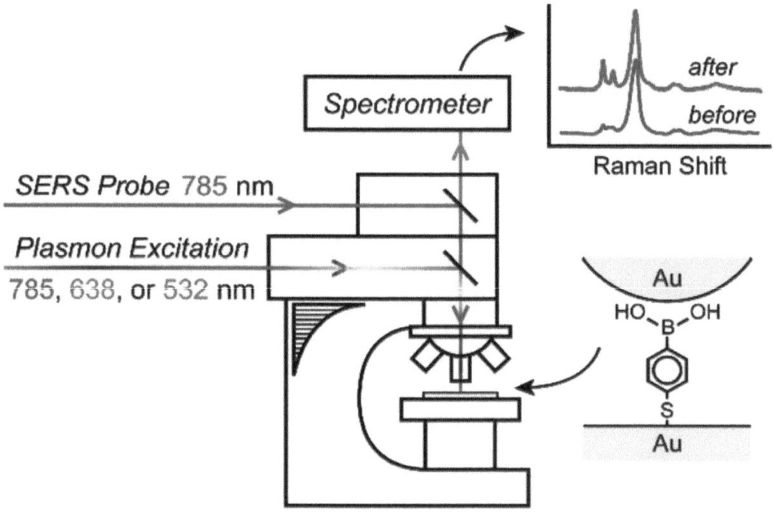

Figure 3.17 Schematic of the experimental setup for the protodeboronation reaction. The sample is prepared by drop-casting AuNPs (51 nm) onto MPBA SAMs on 150 nm thick Au substrates. The sample is placed under a 50× objective in an upright microscope. A laser (785, 638, or 532 nm) is focused on the sample through the objective for plasmon excitation. Adapted from Ref. [97].

Among the different plausible mechanisms upon irradiation, the reaction by plasmonic heating was ruled out since no reaction took place upon heating to 100°C (without laser excitation). The reaction was more likely due to hot electrons and hot holes that directly transfer to MPBA, which subsequently leads to the formation of BT. Additionally, the authors observed that the presence of oxygen molecules promoted the reaction by opening another electron-transfer channel.

3.5.2 Photoconversion of CO_2

Reduced graphene-coated AuNPs (r-GO AuNPs) were proven to be excellent photocatalysts for the photoconversion of CO_2 into formic acid (HCOOH) (>90%) [98]. It was suggested that reduced graphene layers can play a role as efficient electron acceptors and transporters, facilitating the utilization of hot electrons for plasmonic photocatalysts. The femtosecond transient spectroscopic analysis also showed 8.7 times higher transport efficiency of hot plasmonic electrons in r-GO AuNPs compared with AuNPs.

Very recently, Li et al. [99] developed a Bi_2O_{3-x} catalyst that reduced CO_2 to CO under infrared light. They used a low-intensity LED light at 940 nm and was able to reach a catalytic activity of 4.6 μmol/gh. It was also demonstrated that the photocatalytic activity of Bi_2O_{3-x} was due to a linear dependence on light intensity and temperature.

Zhang et al. [100] reported the reduction of CO_2 to CO using molecular hydrogen as the reducing agent and catalyzed by an Ag-ZrO_2 photocatalyst under UV-visible light. The reaction was studied mechanistically and spectroscopically and found to involve the dual roles of light: charge separation and plasmon heating.

3.5.3 Other Chemical Reactions

Combining an aldehyde, an amine, and phenylacetylene at a high temperature led to the formation of propargylamines in good yields (50–95%) using the excitation of the surface plasmons of AuNPs supported on oxide zinc (AuNP@ZnO) (Scheme 3.11) [101]. The products were obtained after 2 h under LED excitation at 530 nm.

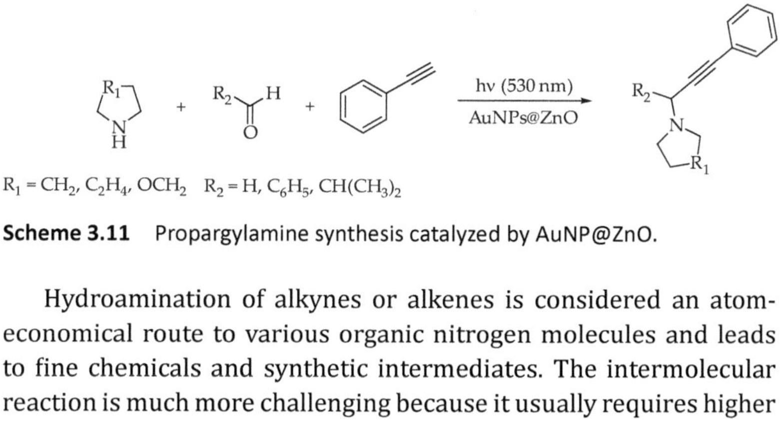

Scheme 3.11 Propargylamine synthesis catalyzed by AuNP@ZnO.

Hydroamination of alkynes or alkenes is considered an atom-economical route to various organic nitrogen molecules and leads to fine chemicals and synthetic intermediates. The intermolecular reaction is much more challenging because it usually requires higher temperatures (generally above 100°C) than the intramolecular version to achieve high yields. However, at high temperatures, the reaction equilibrium shifts toward lower conversions, which creates a dilemma. Thus, Zhao et al. [102] studied the photocatalytic hydroamination of alkynes with aniline molecules using AuNPs supported on oxide nitrogen-doped titanium, AuNP@TiO_2-N, which proved to be highly efficient and selective for the reaction at room temperature under green light irradiation (Scheme 3.12). In this reaction, both AuNPs and the support contribute to the activation of reactants: Aniline molecules preferentially interact with AuNP, while alkynes were adsorbed onto active support sites [102].

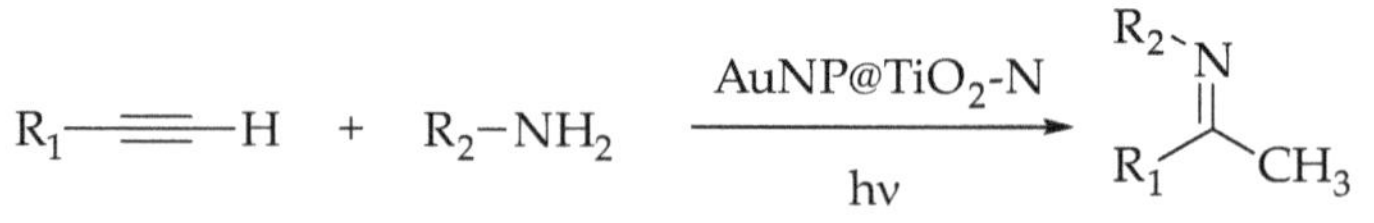

Scheme 3.12 Photocatalytic hydroamination of alkynes catalyzed by AuNP@ TiO_2-N.

The combination of plasmonics and solvent-free chemical synthesis is an attractive green chemistry approach. Meng et al. reported the solvent-free hydroamination of alkynes using plasmonic AuNPs coupled with a TiO_2 2D photonic layer on nanotube arrays (Fig. 3.18) [103]. The solvent-free hydroamination of phenylacetylene using aniline was performed with both Au/ TiO_2 photonic crystal nanotubes (PCNT) and Au/TiO_2 nanotubes. An absence of hydroamination was noted when only the TiO_2 PCNT support was present, confirming that AuNPs are responsible for catalytic activity. The authors suggest that the simultaneous adsorption or activation of the reactants is required for the reaction to proceed with high turnover.

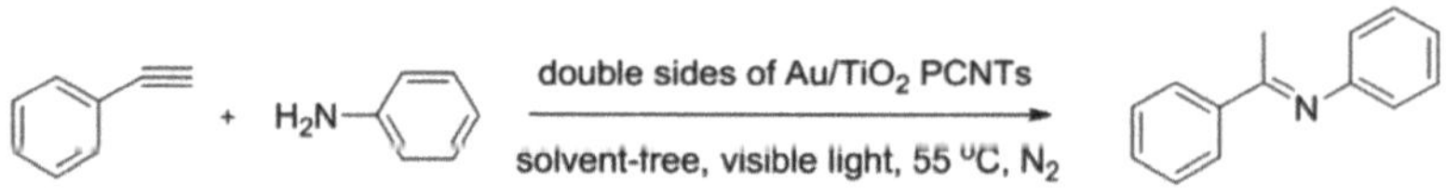

Figure 3.18 Hydroamination of phenylacetylene using AuNP-photonic crystal nanotubes. Adapted from Ref. [103].

The synthesis of 4-benzoylmorpholine from benzaldehyde and morpholine via an amide formation was catalyzed by AuNPs supported on SiO_2 (AuNP@SiO_2) at room temperature and under laser irradiation, and quantitative yields were obtained [104]. This reaction was also extended to a tandem oxidation/amidation process starting with benzyl alcohol and morpholine. In this case, the catalyst irradiation at 532 nm leads to the initial formation of benzaldehyde, through oxidation by hydrogen peroxide present in the middle, and then the amidation reaction took place (Scheme 3.13).

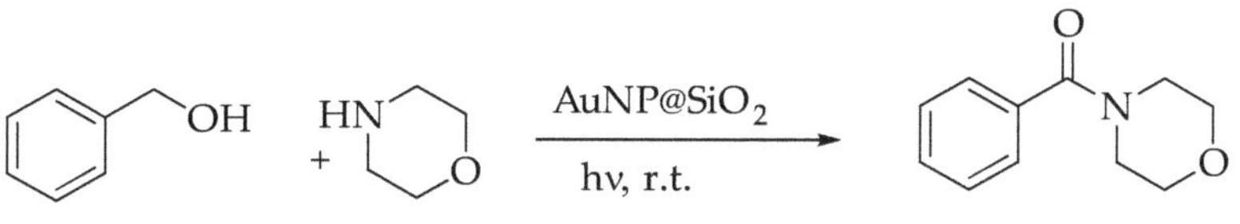

Scheme 3.13 Laser-driven tandem oxidation/amidation reaction catalyzed by $AuNP@SiO_2$.

This reaction was also used in the synthesis of methyl-N-benzylbenzamide (from methyl benzylamine) and *N*-benzylamide (the benzylamine) with high performance and selectivity. A number of other catalysts possessing a plasmon band, such as gold nanorods and gold nanoparticles supported on various oxides, were also investigated; however, low yields were obtained. The highest performing catalyst was $AuNP@SiO_2$, which preserved 90% of its initial activity after a minimum of two reuses.

Bazyar et al. [105] have shown that Au@ZnO core–shell NPs were good photoredox catalysts for the facile trifluoromethylation of a wide range of substrates using sodium trifluoroacetate (Scheme 3.14). The method was extended to the trifluoromethylation of biologically active molecules, and one reaction was even performed at the 10 g scale [105]. For the mechanism, they proposed that gold absorbs visible light, and a hot electron is then transferred to the conduction band (CB) of ZnO, which prevents the electron/hole (e^-/h^+) recombination. This photogenerated h^+ on gold can oxidize CF_3CO_2Na, generating the CF_3 radical through decarboxylation.

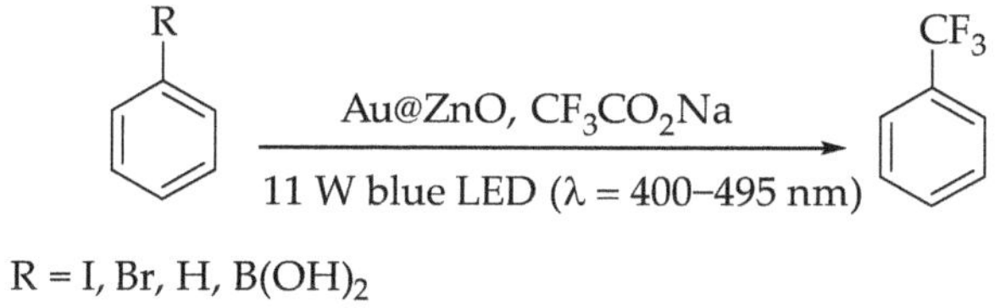

Scheme 3.14 Photocatalytic trifluoromethylation under visible light catalyzed by AuNP@ZnO.

The copper-catalyzed 1,3-dipolar azide–alkyne cycloaddition reaction, known as Huisgen reaction and commonly called "click-chemistry," allows the formation of triazole molecules. The catalytically active species in this reaction is Cu(I), but Cu(I) is usually not stable when heated or exposed to oxygen. Thus, this reaction is usually conducted with Cu(II) species and a reducing agent. Sun et al. [106] studied this reaction using AuNPs and Cu(II) precursors under irradiation, and very good yields were obtained (Scheme 3.15). The observed catalytic enhancement of this reaction was due to the synergistic effect of LSPR inducing a local heating effect, and plasmonic hot electrons are formed, which reduce Cu(II) species to catalytically active Cu(I) species. Au–CuO hybrid nanostructures were described that proved to be an efficient means for coupling visible-light energy, which heterogeneously catalyzes the "click-chemistry" reaction, without the use of any conventional heating and a chemical reducing agent.

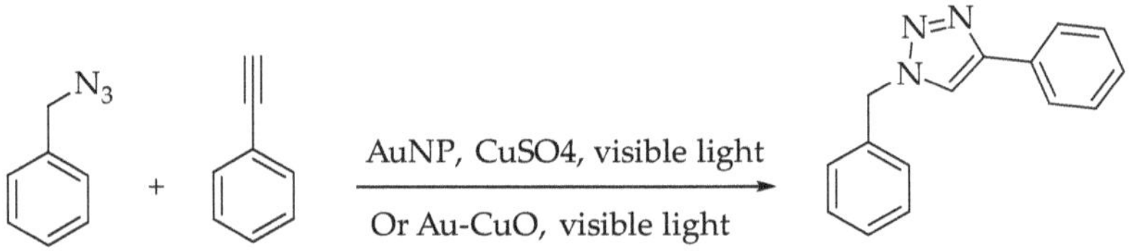

Scheme 3.15 Surface plasmon resonances enhanced copper catalyzed 1,3-dipolar azide–alkyne cycloaddition reaction.

Very recently, Guselnikova et al. [107] studied the influence of temperature on plasmon-driven click-chemistry. The alkyne reagent was grafted on the surface of spherical AuNPs through a diazonium approach, and this nanotemplate was dispersed in acetonitrile and reacted with 4-azidobenzoic acid under light irradiation. When they performed the reaction at −35°C, they noticed an important increase in reactivity, which was unexpected. To explain this phenomenon, they performed DFT calculations and proposed that the decrease in electron–phonon scattering and retardation of organic molecule relaxation after plasmon-induced excitation induced this reversed dependence on the reaction rate.

Thiol-ene coupling is part of the "click-chemistry" set of reactions and consists of the reaction of alkene with a thiol to form an alkyl sulfide. Tijunelyte et al. [108] studied this thiol-ene "click" reaction using gold nanocylinders under visible light. 1,4-butanedithiol, bearing a double-bond terminal group, was initially self-assembled onto the Au cylinders. The modified surface was then exposed to thiophenol in the liquid phase in the presence of a radical initiator (2,2′-azobis(2-methylpropionamidine)dihydrochloride, AAPH) and irradiated at 660 nm (Scheme 3.16). Under irradiation, the reaction rate was accelerated up to 20 times and was interpreted as the combination of both photonic and thermal effects, which can enhance and control chemical reactions at the nanoscale level.

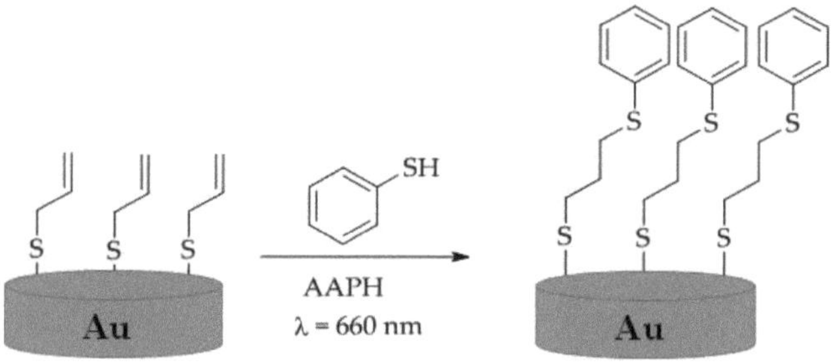

Scheme 3.16 Thio-ene click chemistry reaction taking place at the surface of laser-illuminated Au nanocylinders.

A Friedel–Crafts alkylation was performed using a nanocatalyst composed of AuNPs supported on either commercial niobium oxide HY 340 or mesoporous niobium oxide, these supports replacing the classical Lewis acid (Scheme 3.17) [109]. Under irradiation of this nanocatalyst, anisole and benzyl chloride reacted at 80°C and 100% conversion was obtained within 2 h. In the absence of LED irradiation, for the reaction to proceed, the temperature was increased beyond 120°C and the conversion reached 80% after 2 h, while at 80°C, no conversion was observed.

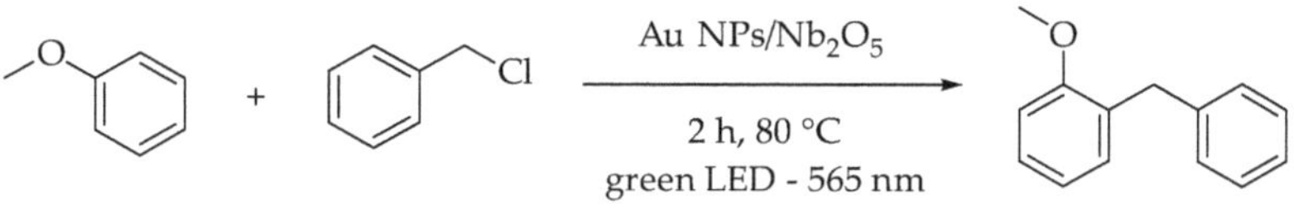

Scheme 3.17 Friedel–Crafts reaction taking place at the surface of laser-illuminated Au NPs/Nb_2O_5.

An interesting tandem oxidative coupling of alcohols and amines using Au/CeO_2 nanorods promoted by visible light was reported by Teixeira et al. [110]. In this reaction, it is proposed that hot electrons generated from the LSPR on AuNPs are trapped by oxygen vacancies in the CeO_2 nanorods, which are then picked up by O_2 molecules on the surface. Concomitantly, hot holes generated by the AuNPs promote α-hydride abstraction from the alcohol, yielding a nascent aldehyde, which in turn couples with an amine (Fig. 3.19).

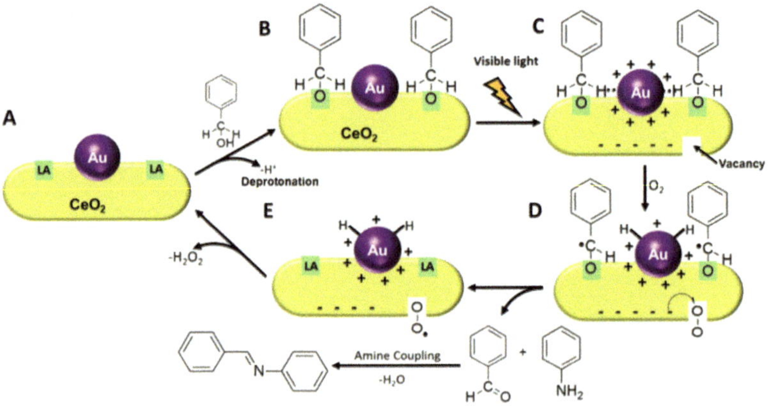

Figure 3.19 Proposed mechanism for Au–CeO_2-mediated coupling of benzyl alcohol with aniline under visible-light irradiation. Adapted from Ref. [110].

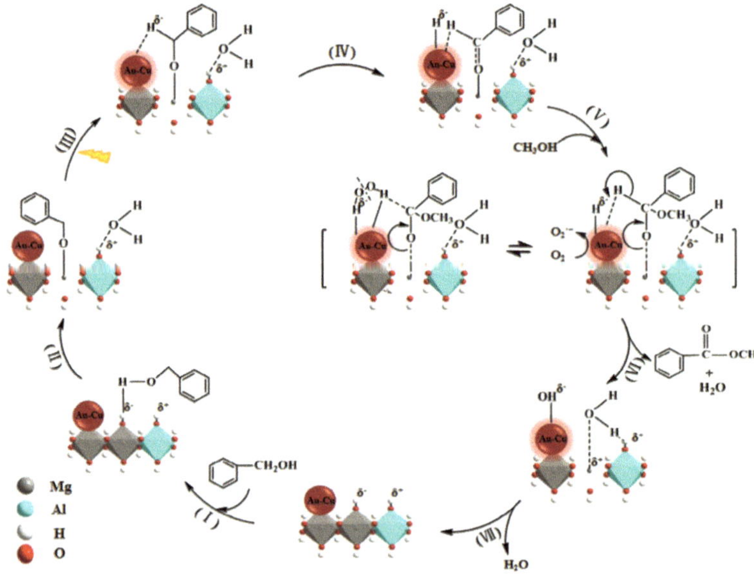

Figure 3.20 Proposed mechanism for the photocatalytic oxidative cross-esterification of methanol and benzyl alcohol. Adapted from Ref. [111].

Alcohols were also found to react with methanol in an LSPR-promoted oxidative esterification reaction reported by Wang et al. [111]. In this chapter, layered double hydroxide Au–Cu NPs catalyzed the esterification of methanol with benzyl alcohol in the presence of molecular O_2. Apparently, the presence of copper in the AuNPs promotes the formation of the superoxide radical $O_2^{\bullet-}$, which is invoked in the mechanism shown in Fig. 3.20.

References

1. Gelle, A., Jin, T., de la Garza, L., Price, G. D., Besteiro, L. V., Moores, A. Applications of plasmon-enhanced nanocatalysis to organic transformations. *Chem. Rev.* 2020, **120**(2), 986–1041. https://doi.org/10.1021/acs.chemrev.9b00187.
2. Gellé, A. and Moores, A. Plasmonic nanoparticles: Photocatalysts with a bright future. *Curr. Opin. Green Sustain. Chem.* 2019, **15**, 60–66. https://doi.org/10.1016/j.cogsc.2018.10.002.
3. Tian, Y. and Tatsuma, T. Mechanisms and applications of plasmon-induced charge separation at TiO_2 films loaded with gold nanoparticles. *J. Am. Chem. Soc.* 2005, **127**(20), 7632–7637. https://doi.org/10.1021/ja042192u.
4. Tanaka, A., Hashimoto, K., and Kominami, H. Selective photocatalytic oxidation of aromatic alcohols to aldehydes in an aqueous suspension of gold nanoparticles supported on cerium(IV) oxide under irradiation of green light. *Chem. Commun.* 2011, **47**(37), 10446–10448. https://doi.org/10.1039/C1CC13801D.
5. Tanaka, A., Hashimoto, K., and Kominami, H. Preparation of Au/CeO_2 exhibiting strong surface plasmon resonance effective for selective or chemoselective oxidation of alcohols to aldehydes or ketones in aqueous suspensions under irradiation by green light. *J. Am. Chem. Soc.* 2012, **134**(35), 14526–14533. https://doi.org/10.1021/ja305225s.
6. Tsukamoto, D., Shiraishi, Y., Sugano, Y., Ichikawa, S., Tanaka, S., and Hirai, T. Gold nanoparticles located at the interface of anatase/rutile TiO_2 particles as active plasmonic photocatalysts for aerobic oxidation. *J. Am. Chem. Soc.* 2012, **134**(14), 6309–6315. https://doi.org/10.1021/ja2120647.
7. Zhai, W., Xue, S., Zhu, A., Luo, Y., and Tian, Y. Plasmon-driven selective oxidation of aromatic alcohols to aldehydes in water with recyclable Pt/TiO_2 nanocomposites. *ChemCatChem* 2011, **3**(1), 127–130. https://doi.org/10.1002/cctc.201000303.

8. Hallett-Tapley, G. L., Silvero, M. J., Gonzalez-Bejar, M., Grenier, M., Netto-Ferreira, J. C., and Scaiano, J. C. Plasmon-mediated catalytic oxidation of sec-phenethyl and benzyl alcohols. *J. Phys. Chem. C* 2011, **115**(21), 10784–10790. https://doi.org/10.1021/jp202769a.
9. Hallett-Tapley, G. L., Silvero, M. J., Bueno-Alejo, C. J., González-Béjar, M., McTiernan, C. D., Grenier, M., Netto-Ferreira, J. C., and Scaiano, J. C. Supported gold nanoparticles as efficient catalysts in the solventless plasmon mediated oxidation of *sec*-phenethyl and benzyl alcohol. *J. Phys. Chem. C* 2013, **117**(23), 12279–12288. https://doi.org/10.1021/jp311069v.
10. Wee, T.-L. E., Schmidt, L. C., and Scaiano, J. C. Photooxidation of 9-anthraldehyde catalyzed by gold nanoparticles: Solution and single nanoparticle studies using fluorescence lifetime imaging. *J. Phys. Chem. C* 2012, **116**(45), 24373–24379. https://doi.org/10.1021/jp308956y.
11. Sarina, S., Zhu, H., Jaatinen, E., Xiao, Q., Liu, H., Jia, J., Chen, C., and Zhao, J. Enhancing catalytic performance of palladium in gold and palladium alloy nanoparticles for organic synthesis reactions through visible light irradiation at ambient temperatures. *J. Am. Chem. Soc.* 2013, **135**(15), 5793–5801. https://doi.org/10.1021/ja400527a.
12. Eerdemutu, Ding, L., Li, C., Zhao, S., and Sheng, X. Synthesis of imines by selective oxidation of amines on alloy nanoparticles of palladium and gold under visible light irradiation at ambient temperatures. *Catal. Lett.* 2020, **150**(6), 1757–1765. https://doi.org/10.1007/s10562-019-03079-w.
13. Swaminathan, S., Rao, V. G., Bera, J. K., and Chandra, M. The pivotal role of hot carriers in plasmonic catalysis of C–N bond forming reaction of amines. *Angew. Chem. Int. Ed.* 2021, **60**(22), 12532–12538. https://doi.org/10.1002/anie.202101639.
14. Zhang, K., Xi, Z., Wu, Z., Lu, G., and Huang, X. Visible-light-induced selective oxidation of amines into imines over UiO-66-NH_2@Au@COF core–shell photocatalysts. *ACS Sustain. Chem. Eng.* 2021, **9**(37), 12623–12633. https://doi.org/10.1021/acssuschemeng.1c04112.
15. Kumar, A., Kumari, N., Dubbu, S., Kumar, S., Kwon, T., Koo, J. H., Lim, J., Kim, I., Cho, Y.-K., Rho, J., and Lee, I. S. Nanocatalosomes as plasmonic bilayer shells with interlayer catalytic nanospaces for solar-light-induced reactions. *Angew. Chem. Int. Ed.* 2020, **59**(24), 9460–9469. https://doi.org/10.1002/anie.202001531.
16. Huang, X., Li, Y., Chen, Y., Zhou, H., Duan, X., and Huang, Y. Plasmonic and catalytic AuPd nanowheels for the efficient conversion of light into chemical energy. *Angew. Chem. Int. Ed.* 2013, **52**(23), 6063–6067. https://doi.org/10.1002/anie.201301096.

17. Jiang, T., Jia, C., Zhang, L., He, S., Sang, Y., Li, H., Li, Y., Xu, X., and Liu, H. Gold and gold–palladium alloy nanoparticles on heterostructured TiO_2 nanobelts as plasmonic photocatalysts for benzyl alcohol oxidation. *Nanoscale* 2014, **7**(1), 209–217. https://doi.org/10.1039/C4NR05905K.

18. Li, H., Qin, F., Yang, Z., Cui, X., Wang, J., and Zhang, L. New reaction pathway induced by plasmon for selective benzyl alcohol oxidation on BiOCl possessing oxygen vacancies. *J. Am. Chem. Soc.* 2017, **139**(9), 3513–3521. https://doi.org/10.1021/jacs.6b12850.

19. Liu, L., Ouyang, S., and Ye, J. Gold-nanorod-photosensitized titanium dioxide with wide-range visible-light harvesting based on localized surface plasmon resonance. *Angew. Chem. Int. Ed.* 2013, **52**(26), 6689–6693. https://doi.org/10.1002/anie.201300239.

20. Nadeem, M. A. and Idriss, H. Photo-thermal reactions of ethanol over Ag/TiO_2 catalysts. The role of silver plasmon resonance in the reaction kinetics. *Chem. Commun.* 2018, **54**(41), 5197–5200. https://doi.org/10.1039/C8CC01814F.

21. Kowalska, E., Mahaney, O. O. P., Abe, R., and Ohtani, B. Visible-light-induced photocatalysis through surface plasmon excitation of gold on titania surfaces. *Phys. Chem. Chem. Phys.* 2010, **12**(10), 2344–2355. https://doi.org/10.1039/B917399D.

22. Kowalska, E., Abe, R., and Ohtani, B. Visible light-induced photocatalytic reaction of gold-modified titanium(IV) oxide particles: Action spectrum analysis. *Chem. Commun.* 2009, 2, 241–243. https://doi.org/10.1039/B815679D.

23. Sugano, Y., Shiraishi, Y., Tsukamoto, D., Ichikawa, S., Tanaka, S., and Hirai, T. Supported Au–Cu bimetallic alloy nanoparticles: An aerobic oxidation catalyst with regenerable activity by visible-light irradiation. *Angew. Chem. Int. Ed.* 2013, **52**(20), 5295–5299. https://doi.org/10.1002/anie.201301669.

24. Colmenares, J. C., Lisowski, P., Łomot, D., Chernyayeva, O., and Lisovytskiy, D. Sonophotodeposition of bimetallic photocatalysts $Pd–Au/TiO_2$: Application to selective oxidation of methanol to methyl formate. *ChemSusChem* 2015, **8**(10), 1676–1685. https://doi.org/10.1002/cssc.201403125.

25. Sun, H., Chen, S., Yang, W., Wang, L., Tang, R., Zhang, X., Zheng, R., Gu, S., Jiang, Y., Liang, W., and Huang, J. Plasmon-enhanced alcohol oxidations over porous carbon nanosphere-supported palladium and gold bimetallic nanocatalyst. *Appl. Catal. B Environ.* 2021, **292**, 120151. https://doi.org/10.1016/j.apcatb.2021.120151.

26. Wang, Z., Du, J., Zhang, Y., Han, J., Huang, S., Hirata, A., and Chen, M. Free-standing nanoporous gold for direct plasmon enhanced electro-oxidation of alcohol molecules. *Nano Energy* 2019, **56**, 286–293. https://doi.org/10.1016/j.nanoen.2018.11.060.

27. Zhu, P., Sun, X., Wang, Y., Zhang, J., Gu, X., and Zheng, Z. Multifunctional oxygen vacancies in WO_{3-x} for catalytic alkylation of C–H by alcohols under red-light. *J. Catal.* 2021, **402**, 208–217. https://doi.org/10.1016/j.jcat.2021.08.040.

28. Abad, A., Corma, A., and García, H. Supported gold nanoparticles for aerobic, solventless oxidation of allylic alcohols. *Pure Appl. Chem.* 2007, **79**(11), 1847–1854. https://doi.org/10.1351/pac200779111847.

29. Landry, M. J., Gellé, A., Meng, B. Y., Barrett, C. J., and Moores, A. Surface-plasmon-mediated hydrogenation of carbonyls catalyzed by silver nanocubes under visible light. *ACS Catal.* 2017, **7**(9), 6128–6133. https://doi.org/10.1021/acscatal.7b02128.

30. Christopher, P., Xin, H., and Linic, S. Visible-light-enhanced catalytic oxidation reactions on plasmonic silver nanostructures. *Nat. Chem.* 2011, **3**(6), 467–472. https://doi.org/10.1038/nchem.1032.

31. Marimuthu, A., Zhang, J., and Linic, S. Tuning selectivity in propylene epoxidation by plasmon mediated photo-switching of cu oxidation state. *Science* 2013, **339**(6127), 1590–1593. https://doi.org/10.1126/science.1231631.

32. Zhang, X., Kumari, G., Heo, J., and Jain, P. K. In situ formation of catalytically active graphene in ethylene photo-epoxidation. *Nat. Commun.* 2018, **9**(1), 3056. https://doi.org/10.1038/s41467-018-05352-9.

33. Ide, Y., Matsuoka, M., and Ogawa, M. Efficient visible-light-induced photocatalytic activity on gold-nanoparticle-supported layered titanate. *J. Am. Chem. Soc.* 2010, **132**(47), 16762–16764. https://doi.org/10.1021/ja1083514.

34. Zheng, Z., Huang, B., Qin, X., Zhang, X., Dai, Y., and Whangbo, M.-H. Facile in situ synthesis of visible-light plasmonic photocatalysts M@TiO_2 (M = Au, Pt, Ag) and evaluation of their photocatalytic oxidation of benzene to phenol. *J. Mater. Chem.* 2011, **21**(25), 9079–9087. https://doi.org/10.1039/C1JM10983A.

35. Xiao, Y., Liu, J., Mai, J., Pan, C., Cai, X., and Fang, Y. High-performance silver nanoparticles coupled with monolayer hydrated tungsten oxide nanosheets: The structural effects in photocatalytic oxidation of cyclohexane. *J. Colloid Interface Sci.* 2018, **516**, 172–181. https://doi.org/10.1016/j.jcis.2018.01.057.

36. Takeuchi, Y., Fujita, T., and Takeyasu, N. Plasmon-mediated chemical transformation from alkane to alkene on a silver nanoparticle array under 532 Nm excitation. *Phys. Chem. Chem. Phys.* 2019, **21**(14), 7502–7507. https://doi.org/10.1039/C9CP00129H.

37. Lee, S. W., Lee, C., Goddeti, K. C., Kim, S. M., and Park, J. Y. Surface plasmon-driven catalytic reactions on a patterned Co_3O_4/Au inverse catalyst. *RSC Adv.* 2017, **7**(88), 56073–56080. https://doi.org/10.1039/C7RA10450B.

38. Lee, S. W., Hong, J. W., Lee, H., Wi, D. H., Kim, S. M., Han, S. W., and Park, J. Y. The surface plasmon-induced hot carrier effect on the catalytic activity of CO oxidation on a Cu_2O/hexoctahedral Au inverse catalyst. *Nanoscale* 2018, **10**(23), 10835–10843. https://doi.org/10.1039/C8NR00555A.

39. Zhang, Y., Li, Q., Liu, C., Shan, X., Chen, X., Dai, W., and Fu, X. The promoted effect of a metal-organic frameworks (ZIF-8) on Au/TiO_2 for CO oxidation at room temperature both in dark and under visible light irradiation. *Appl. Catal. B Environ.* 2018, **224**, 283–294. https://doi.org/10.1016/j.apcatb.2017.10.027.

40. Dong, B., Fang, Y., Xia, L., Xu, H., and Sun, M. Is 4-nitrobenzenethiol converted to p,p′-dimercaptoazobenzene or 4-aminothiophenol by surface photochemistry reaction? *J. Raman Spectrosc.* 2011, **42**(6), 1205–1206. https://doi.org/10.1002/jrs.2937.

41. Dong, B., Fang, Y., Chen, X., Xu, H., and Sun, M. Substrate-, wavelength-, and time-dependent plasmon-assisted surface catalysis reaction of 4-nitrobenzenethiol dimerizing to p,p′-dimercaptoazobenzene on Au, Ag, and Cu films. *Langmuir* 2011, **27**(17), 10677–10682. https://doi.org/10.1021/la2018538.

42. Ren, X., Cao, E., Lin, W., Song, Y., Liang, W., and Wang, J. Recent advances in surface plasmon-driven catalytic reactions. *RSC Adv.* 2017, **7**(50), 31189–31203. https://doi.org/10.1039/C7RA05346K.

43. Zhao, L.-B., Zhang, M., Huang, Y.-F., Williams, C. T., Wu, D.-Y., Ren, B., and Tian, Z.-Q. Theoretical study of plasmon-enhanced surface catalytic coupling reactions of aromatic amines and nitro compounds. *J. Phys. Chem. Lett.* 2014, **5**(7), 1259–1266. https://doi.org/10.1021/jz5003346.

44. Zhu, H., Ke, X., Yang, X., Sarina, S., and Liu, H. Reduction of nitroaromatic compounds on supported gold nanoparticles by visible and ultraviolet light. *Angew. Chem. Int. Ed.* 2010, **49**(50), 9657–9661. https://doi.org/10.1002/anie.201003908.

45. Liu, Z., Huang, Y., Xiao, Q., and Zhu, H. Selective reduction of nitroaromatics to azoxy compounds on supported Ag–Cu alloy nanoparticles through

visible light irradiation. *Green Chem.* 2016, **18**(3), 817–825. https://doi.org/10.1039/C5GC01726B.

46. Liu, Q., Zhang, J., Xing, F., Cheng, C., Wu, Y., and Huang, C. Plasmon-enhanced and controllable synthesis of azobenzene and hydrazobenzene using Au/TiO_2 composite. *Appl. Surf. Sci.* 2020, **500**, 144214. https://doi.org/10.1016/j.apsusc.2019.144214.
47. Ding, Q., Chen, M., Fang, Y., Zhang, Z., and Sun, M. Plasmon-driven diazo coupling reactions of p-nitroaniline via $-NH_2$ or $-NO_2$ in atmosphere environment. *J. Phys. Chem. C* 2017, **121**(9), 5225–5231. https://doi.org/10.1021/acs.jpcc.6b12589.
48. Ke, X., Zhang, X., Zhao, J., Sarina, S., Barry, J., and Zhu, H. Selective reductions using visible light photocatalysts of supported gold nanoparticles. *Green Chem.* 2012, **15**(1), 236–244. https://doi.org/10.1039/C2GC36542A.
49. Kang, L., Xu, P., Zhang, B., Tsai, H., Han, X., and Wang, H.-L. Laser wavelength- and power-dependent plasmon-driven chemical reactions monitored using single particle surface enhanced Raman spectroscopy. *Chem. Commun.* 2013, **49**(33), 3389–3391. https://doi.org/10.1039/C3CC40732B.
50. Kaur, G., Tanwar, S., Kaur, V., Biswas, R., Saini, S., Haldar, K. K., and Sen, T. Interfacial design of gold/silver core–shell nanostars for plasmon-enhanced photocatalytic coupling of 4-aminothiophenol. *J. Mater. Chem. C* 2021, **9**(42), 15284–15294. https://doi.org/10.1039/D1TC03733A.
51. Koopman, W., Sarhan, R. M., Stete, F., Schmitt, C. N. Z., and Bargheer, M. Decoding the kinetic limitations of plasmon catalysis: The case of 4-nitrothiophenol dimerization. *Nanoscale* 2020, **12**(48), 24411–24418. https://doi.org/10.1039/D0NR06039A
52. Cho, F.-H., Kuo, S.-C., and Lai, Y.-H. Surface-plasmon-induced azo coupling reaction between nitro compounds on dendritic silver monitored by surface-enhanced raman spectroscopy. *RSC Adv.* 2017, **7**(17), 10259–10265. https://doi.org/10.1039/C7RA00374A.
53. Hartman, T., Wondergem, C. S., Kumar, N., van den Berg, A., and Weckhuysen, B. M. Surface- and tip-enhanced raman spectroscopy in catalysis. *J. Phys. Chem. Lett.* 2016, **7**(8), 1570–1584. https://doi.org/10.1021/acs.jpclett.6b00147.
54. Zhang, Z., Kinzel, D., and Deckert, V. Photo-induced or plasmon-induced reaction: Investigation of the light-induced azo-coupling of amino groups. *J. Phys. Chem. C* 2016, **120**(37), 20978–20983. https://doi.org/10.1021/acs.jpcc.6b03233.

55. Huang, Y.-F., Wu, D.-Y., Zhu, H.-P., Zhao, L.-B., Liu, G.-K., Ren, B., and Tian, Z.-Q. Surface-enhanced raman spectroscopic study of p-aminothiophenol. *Phys. Chem. Chem. Phys.* 2012, **14**(24), 8485–8497. https://doi.org/10.1039/C2CP40558J.

56. Ren, X., Tan, E., Lang, X., You, T., Jiang, L., Zhang, H., Yin, P., and Guo, L. Observing reduction of 4-nitrobenzenthiol on gold nanoparticles in situ using surface-enhanced Raman spectroscopy. *Phys. Chem. Chem. Phys.* 2013, **15**(34), 14196–14201. https://doi.org/10.1039/C3CP51385H.

57. Pradhan, N., Pal, A., and Pal, T. Silver nanoparticle catalyzed reduction of aromatic nitro compounds. *Colloids Surf. Physicochem. Eng. Asp.* 2002, **196**(2), 247–257. https://doi.org/10.1016/S0927-7757(01)01040-8.

58. Wu, W., Lei, M., Yang, S., Zhou, L., Liu, L., Xiao, X., Jiang, C., and Roy, V. A. L. A one-pot route to the synthesis of alloyed Cu/Ag bimetallic nanoparticles with different mass ratios for catalytic reduction of 4-nitrophenol. *J. Mater. Chem. A* 2015, **3**(7), 3450–3455. https://doi.org/10.1039/C4TA06567K.

59. Han, Q., Zhang, C., Gao, W., Han, Z., Liu, T., Li, C., Wang, Z., He, E., and Zheng, H. Ag-Au alloy nanoparticles: Synthesis and in situ monitoring SERS of plasmonic catalysis. *Sens. Actuators B-Chem.* 2016, **231**, 609–614. https://doi.org/10.1016/j.snb.2016.03.068.

60. Shah, D. and Kaur, H. Resin-trapped gold nanoparticles: An efficient catalyst for reduction of nitro compounds and Suzuki–Miyaura coupling. *J. Mol. Catal. Chem.* 2014, **381**, 70–76. https://doi.org/10.1016/j.molcata.2013.10.004.

61. Liu, H., Yu, J., Liu, Y., and Liu, Y. Plasmonic acceleration of nitrophenol reduction upon catalysis by robust gold nanoparticle donut assemblies. *Catal. Commun.* 2020, **146**, 106137. https://doi.org/10.1016/j.catcom.2020.106137.

62. Zeng, J., Zhang, Q., Chen, J., and Xia, Y. A comparison study of the catalytic properties of Au-based nanocages, nanoboxes, and nanoparticles. *Nano Lett.* 2010, **10**(1), 30–35. https://doi.org/10.1021/nl903062e.

63. Wu, X.-Q., Wu, X.-W., Huang, Q., Shen, J.-S., and Zhang, H.-W. In situ synthesized gold nanoparticles in hydrogels for catalytic reduction of nitroaromatic compounds. *Appl. Surf. Sci.* 2015, **331**, 210–218. https://doi.org/10.1016/j.apsusc.2015.01.077.

64. Cui, Q., Xia, B., Mitzscherling, S., Masic, A., Li, L., Bargheer, M., and Möhwald, H. Preparation of gold nanostars and their study in selective catalytic reactions. *Colloids Surf. Physicochem. Eng. Asp.* 2015, **465**, 20–25. https://doi.org/10.1016/j.colsurfa.2014.10.028.

65. Haldar, K. K., Kundu, S., and Patra, A. Core-size-dependent catalytic properties of bimetallic Au/Ag core–shell nanoparticles. *ACS Appl. Mater. Interfaces* 2014, **6**(24), 21946–21953. https://doi.org/10.1021/am507391d.

66. Xiao, Q., Sarina, S., Waclawik, E. R., Jia, J., Chang, J., Riches, J. D., Wu, H., Zheng, Z., and Zhu, H. Alloying gold with copper makes for a highly selective visible-light photocatalyst for the reduction of nitroaromatics to anilines. *ACS Catal.* 2016, **6**(3), 1744–1753. https://doi.org/10.1021/acscatal.5b02643.

67. Hajfathalian, M., Gilroy, K. D., Yaghoubzade, A., Sundar, A., Tan, T., Hughes, R. A., and Neretina, S. Photocatalytic enhancements to the reduction of 4-nitrophenol by resonantly excited triangular gold–copper nanostructures. *J. Phys. Chem. C* 2015, **119**(30), 17308–17315. https://doi.org/10.1021/acs.jpcc.5b04618.

68. He, R., Wang, Y.-C., Wang, X., Wang, Z., Liu, G., Zhou, W., Wen, L., Li, Q., Wang, X., Chen, X., Zeng, J., and Hou, J. G. Facile synthesis of pentacle gold–copper alloy nanocrystals and their plasmonic and catalytic properties. *Nat. Commun.* 2014, **5**(1), 4327. https://doi.org/10.1038/ncomms5327.

69. Vysakh, A. B., Babu, C. L., and Vinod, C. P. Demonstration of synergistic catalysis in Au@Ni bimetallic core–shell nanostructures. *J. Phys. Chem. C* 2015, **119**(15), 8138–8146. https://doi.org/10.1021/jp5128089.

70. Li, J., Long, Y., Liu, Y., Zhang, L., Wang, Q., Wang, X., Song, S., and Zhang, H. Robust synthesis of gold-based multishell structures as plasmonic catalysts for selective hydrogenation of 4-nitrostyrene. *Angew. Chem. Int. Ed.* 2020, **59**(3), 1103–1107. https://doi.org/10.1002/anie.201910836.

71. Verma, A., Shukla, M., Kumar, S., Pal, S., and Sinha, I. Mechanism of visible light enhanced catalysis over curcumin functionalized Ag nanocatalysts. *Spectrochim. Acta. A. Mol. Biomol. Spectrosc.* 2020, **240**, 118534. https://doi.org/10.1016/j.saa.2020.118534.

72. Barbosa, E. C. M., Fiorio, J. L., Mou, T., Wang, B., Rossi, L. M., and Camargo, P. H. C. Reaction pathway dependence in plasmonic catalysis: Hydrogenation as a model molecular transformation. *Chem. – Eur. J.* 2018, **24**(47), 12330–12339. https://doi.org/10.1002/chem.201705749.

73. Yin, H., Kuwahara, Y., Mori, K., and Yamashita, H. RuPd alloy nanoparticles supported on plasmonic H_xMoO_{3-y} for efficient photocatalytic reduction of p-nitrophenol. *Eur. J. Inorg. Chem.* 2019, **2019**(33), 3745–3752. https://doi.org/10.1002/ejic.201900801.

74. Tong, F., Liang, X., Ma, F., Bao, X., Wang, Z., Liu, Y., Wang, P., Cheng, H., Dai, Y., Huang, B., and Zheng, Z. Plasmon-mediated nitrobenzene hydrogenation with formate as the hydrogen donor studied at a single-particle level. *ACS Catal.* 2021, **11**(7), 3801–3809. https://doi.org/10.1021/acscatal.1c00164.

75. Hao, C.-H., Guo, X.-N., Pan, Y.-T., Chen, S., Jiao, Z.-F., Yang, H., and Guo, X.-Y. Visible-light-driven selective photocatalytic hydrogenation of cinnamaldehyde over Au/SiC catalysts. *J. Am. Chem. Soc.* 2016, **138**(30), 9361–9364. https://doi.org/10.1021/jacs.6b04175.

76. Alejo, C. J. B., Fasciani, C., Grenier, M., Netto-Ferreira, J. C., and Scaiano, J. C. Reduction of resazurin to resorufin catalyzed by gold nanoparticles: Dramatic reaction acceleration by laser or LED plasmon excitation. *Catal. Sci. Technol.* 2011, **1**(8), 1506–1511. https://doi.org/10.1039/c1cy00236h.

77. Xu, W., Kong, J. S., Yeh, Y.-T. E., and Chen, P. Single-molecule nanocatalysis reveals heterogeneous reaction pathways and catalytic dynamics. *Nat. Mater.* 2008, **7**(12), 992–996. https://doi.org/10.1038/nmat2319.

78. Zhou, X., Xu, W., Liu, G., Panda, D., and Chen, P. Size-dependent catalytic activity and dynamics of gold nanoparticles at the single-molecule level. *J. Am. Chem. Soc.* 2010, **132**(1), 138–146. https://doi.org/10.1021/ja904307n.

79. Chen, P., Zhou, X., Andoy, N. M., Han, K.-S., Choudhary, E., Zou, N., Chen, G., and Shen, H. Spatiotemporal catalytic dynamics within single nanocatalysts revealed by single-molecule microscopy. *Chem. Soc. Rev.* 2014, **43**(4), 1107–1117. https://doi.org/10.1039/C3CS60215J.

80. Quiroz, J., Barbosa, E. C. M., Araujo, T. P., Fiorio, J. L., Wang, Y.-C., Zou, Y.-C., Mou, T., Alves, T. V., de Oliveira, D. C., Wang, B., Haigh, S. J., Rossi, L. M., and Camargo, P. H. C. Controlling reaction selectivity over hybrid plasmonic nanocatalysts. *Nano Lett.* 2018, **18**(11), 7289–7297. https://doi.org/10.1021/acs.nanolett.8b03499.

81. Wang, R., Liu, H., Wang, X., Li, X., Gu, X., and Zheng, Z. Plasmon-enhanced furfural hydrogenation catalyzed by stable carbon-coated copper nanoparticles driven from metal–organic frameworks. *Catal. Sci. Technol.* 2020, **10**(19), 6483–6494. https://doi.org/10.1039/D0CY01162B.

82. Garcia-Garcia, I., Lovell, E. C., Wong, R. J., Barrio, V. L., Scott, J., Cambra, J. F., and Amal, R. Silver-based plasmonic catalysts for carbon dioxide reduction. *ACS Sustain. Chem. Eng.* 2020, **8**(4), 1879–1887. https://doi.org/10.1021/acssuschemeng.9b06146.

83. Fasciani, C., Alejo, C. J. B., Grenier, M., Netto-Ferreira, J. C., and Scaiano, J. C. High-temperature organic reactions at room temperature using plasmon excitation: Decomposition of dicumyl peroxide. *Org. Lett.* 2011, **13**(2), 204–207. https://doi.org/10.1021/ol1026427.

84. Bakhtiari, A. B. S., Hsiao, D., Jin, G., Gates, B. D., and Branda, N. R. An efficient method based on the photothermal effect for the release of molecules from metal nanoparticle surfaces. *Angew. Chem.-Int. Ed.* 2009, **48**(23), 4166–4169. https://doi.org/10.1002/anie.200805303.

85. Liang, C., Zhang, Y., Zhang, B., Liu, X.-M., Gao, G.-L., Cao, J., and Xu, P. Plasmonic heating-promoted photothermal synthesis of α-cyanoacrylonitriles over Au/h-BN catalysts. *Front. Chem.* 2021, **9**.

86. Han, D., Bao, Z., Xing, H., Yang, Y., Ren, Q., and Zhang, Z. Fabrication of plasmonic Au–Pd alloy nanoparticles for photocatalytic Suzuki–Miyaura reactions under ambient conditions. *Nanoscale* 2017, **9**(18), 6026–6032. https://doi.org/10.1039/C7NR01950E.

87. Wang, F., Li, C., Chen, H., Jiang, R., Sun, L.-D., Li, Q., Wang, J., Yu, J. C., and Yan, C.-H. Plasmonic harvesting of light energy for Suzuki coupling reactions. *J. Am. Chem. Soc.* 2013, **135**(15), 5588–5601. https://doi.org/10.1021/ja310501y.

88. Yoshii, T., Kuwahara, Y., Mori, K., and Yamashita, H. Design of Pd–graphene–Au nanorod nanocomposite catalyst for boosting Suzuki–Miyaura coupling reaction by assistance of surface plasmon resonance. *J. Phys. Chem. C* 2019, **123**(40), 24575–24583. https://doi.org/10.1021/acs.jpcc.9b06609.

89. Zhu, X., Jia, H., Zhu, X.-M., Cheng, S., Zhuo, X., Qin, F., Yang, Z., and Wang, J. Selective Pd deposition on Au nanobipyramids and Pd site-dependent plasmonic photocatalytic activity. *Adv. Funct. Mater.* 2017, **27**(22), 1700016. https://doi.org/10.1002/adfm.201700016.

90. Sarhid, I., Abdellah, I., Martini, C., Huc, V., Dragoe, D., Beaunier, P., Lampre, I., and Remita, H. Plasmonic catalysis for the Suzuki–Miyaura cross-coupling reaction using palladium nanoflowers. *New J. Chem.* 2019, **43**(11), 4349–4355. https://doi.org/10.1039/C8NJ06370B.

91. Koohgard, M. and Hosseini-Sarvari, M. Enhancement of Suzuki–Miyaura coupling reaction by photocatalytic palladium nanoparticles anchored to TiO_2 under visible light irradiation. *Catal. Commun.* 2018, **111**, 10–15. https://doi.org/10.1016/j.catcom.2018.03.026.

92. Gangishetty, M. K., Fontes, A. M., Malta, M., Kelly, T. L., and Scott, R. W. J. Improving the rates of Pd-catalyzed reactions by exciting the surface plasmons of AuPd bimetallic nanotriangles. *RSC Adv.* 2017, **7**(64), 40218–40226. https://doi.org/10.1039/C7RA07264C.

93. Nemygina, N. A., Nikoshvili, L. Zh., Tiamina, I. Yu., Bykov, A. V., Smirnov, I. S., LaGrange, T., Kaszkur, Z., Matveeva, V. G., Sulman, E. M., and Kiwi-Minsker, L. Au core–Pd shell bimetallic nanoparticles immobilized within hyper-cross-linked polystyrene for mechanistic study of Suzuki cross-coupling: Homogeneous or heterogeneous catalysis? *Org. Process Res. Dev.* 2018, **22**(12), 1606–1613. https://doi.org/10.1021/acs.oprd.8b00272.

94. Rohani, S., Ziarati, A., Ziarani, G. M., Badiei, A., and Burgi, T. Engineering of highly active Au/Pd supported on hydrogenated urchin-like yolk@shell TiO_2 for visible light photocatalytic Suzuki coupling. *Catal. Sci. Technol.* 2019, **9**(14), 3820–3827. https://doi.org/10.1039/C9CY00798A.

95. Li, Z. and Kurouski, D. Probing the plasmon-driven Suzuki–Miyaura coupling reactions with cargo-TERS towards tailored catalysis. *Nanoscale* 2021, **13**(27), 11793–11799. https://doi.org/10.1039/D1NR02478G.

96. Li, Z., Gong, Y., Zhang, X., Wen, Y., Yao, J., Hu, M., He, M., Liu, J., Li, R., Wang, F., and Zhang, C. Plasmonic coupling-enhanced in situ photothermal nanoreactor with shape selective catalysis for C–C coupling reaction. *Nano Res.* 2020, **13**(10), 2812–2818. https://doi.org/10.1007/s12274-020-2933-3.

97. Huynh, L. T. M., Trinh, H. D., Lee, S., and Yoon, S. Plasmon-driven protodeboronation reactions in nanogaps. *Nanoscale* 2020. https://doi.org/10.1039/D0NR07023H.

98. Kumar, D., Lee, A., Lee, T., Lim, M., and Lim, D.-K. Ultrafast and efficient transport of hot plasmonic electrons by graphene for Pt free, highly efficient visible-light responsive photocatalyst. *Nano Lett.* 2016, **16**(3), 1760–1767. https://doi.org/10.1021/acs.nanolett.5b04764.

99. Li, Y., Wen, M., Wang, Y., Tian, G., Wang, C., and Zhao, J. Plasmonic hot electrons from oxygen vacancies for infrared light-driven catalytic CO_2 reduction on Bi_2O_{3-x}. *Angew. Chem. Int. Ed.* https://doi.org/10.1002/anie.202010156.

100. Zhang, H., Itoi, T., Konishi, T., and Izumi, Y. Dual photocatalytic roles of light: Charge separation at the band gap and heat via localized surface plasmon resonance to convert CO_2 into CO over silver-zirconium oxide. *J. Am. Chem. Soc.* 2019, **141**(15), 6292–6301. https://doi.org/10.1021/jacs.8b13894.

101. González-Béjar, M., Peters, K., Hallett-Tapley, G. L., Grenier, M., and Scaiano, J. C. Rapid one-pot propargylamine synthesis by plasmon

mediated catalysis with gold nanoparticles on ZnO under ambient conditions. *Chem. Commun.* 2013, **49**(17), 1732–1734. https://doi.org/10.1039/C3CC38287G.

102. Zhao, J., Zheng, Z., Bottle, S., Chou, A., Sarina, S., and Zhu, H. Highly efficient and selective photocatalytic hydroamination of alkynes by supported gold nanoparticles using visible light at ambient temperature. *Chem. Commun.* 2013, **49**(26), 2676–2678. https://doi.org/10.1039/C3CC38985E.

103. Meng, J., Xiao, Y., Huang, Y., Liu, X., and Zhao, J. Solvent-free hydroamination of phenylacetylene by plasmonic gold nanoparticles coupled with a TiO_2 2D photonic layer on nanotube arrays. *New J. Chem.* 2020, **44**(40), 17289–17295. https://doi.org/10.1039/D0NJ04132G.

104. Pineda, A., Gomez, L., Balu, A. M., Sebastian, V., Ojeda, M., Arruebo, M., Romero, A. A., Santamaria, J., and Luque, R. Laser-driven heterogeneous catalysis: Efficient amide formation catalysed by Au/SiO_2 systems. *Green Chem.* 2013, **15**(8), 2043–2049. https://doi.org/10.1039/C3GC40166A.

105. Bazyar, Z. and Hosseini-Sarvari, M. Au@ZnO core–shell: Scalable photocatalytic trifluoromethylation using CF_3CO_2Na as an inexpensive reagent under visible light irradiation. *Org. Process Res. Dev.* 2019, **23**(11), 2345–2353. https://doi.org/10.1021/acs.oprd.9b00225.

106. Sun, X., Zou, Y., and Jiang, J. Surface plasmon resonances enhanced click chemistry through synergistic photothermal and hot electron effects. *Chem. Commun.* 2019, **55**(33), 4813–4816. https://doi.org/10.1039/C9CC01456J.

107. Guselnikova, O., Váňa, J., Phuong, L. T., Panov, I., Rulíšek, L., Trelin, A., Postnikov, P., Švorčík, V., Andris, E., and Lyutakov, O. Plasmon-assisted click chemistry at low temperature: An inverse temperature effect on the reaction rate. *Chem. Sci.* 2021, **12**(15), 5591–5598. https://doi.org/10.1039/D0SC05898J.

108. Tijunelyte, I., Guenin, E., Lidgi-Guigui, N., Colas, F., Ibrahim, J., Toury, T., and de la Chapelle, M. L. Nanoplasmonics tuned "click chemistry." *Nanoscale* 2016, **8**(13), 7105–7112. https://doi.org/10.1039/C5NR09018K.

109. Dos Santos, C. G., Marquez, D. T., Crites, C.-O. L., Netto-Ferreira, J. C., and Scaiano, J. C. Plasmon heating mediated Friedel-Crafts alkylation of anisole using supported AuNP@Nb_2O_5 catalysts. *Tetrahedron Lett.* 2017, **58**(5), 427–431. https://doi.org/10.1016/j.tetlet.2016.12.048.

110. Teixeira, I. F., Homsi, M. S., Geonmonond, R. S., Rocha, G. F. S. R., Peng, Y.-K., Silva, I. F., Quiroz, J., and Camargo, P. H. C. Hot electrons, hot holes, or both? Tandem synthesis of imines driven by the plasmonic excitation in Au/CeO_2 nanorods. *Nanomaterials* 2020, **10**(8), 1530. https://doi.org/10.3390/nano10081530.

111. Wang, X., Wang, R., Wang, J., Fan, C., and Zheng, Z. The synergistic role of the support surface and Au–Cu alloys in a plasmonic Au–Cu@LDH photocatalyst for the oxidative esterification of benzyl alcohol with methanol. *Phys. Chem. Chem. Phys.* 2020, **22**(3), 1655–1664. https://doi.org/10.1039/C9CP05992.

Chapter 4

Plasmon-Mediated Homogeneous Catalysis

Léa Gimeno,[a] Somayeh Talebzadeh,[b] Scott Trammell,[c] Clémence Queffélec,[a] and D. Andrew Knight[c]
[a]*University of Nantes, CNRS, CEISAM, UMR 6230, F-44000 Nantes, France*
[b]*Florida Institute of Technology, 150 West University Blvd, Melbourne, FL 32901, USA*
[c]*US Naval Research Laboratory, 4555 Overlook Avenue, Washington DC 20375, USA*
aknight@fit.edu

4.1 Introduction

Homogeneous catalytic reactions, especially those involving transition metals and organometallic reagents, have been described in the academic literature and used in industrial-scale reactions for over 60 years. Classic examples include the Wacker process for the oxidation of ethylene to acetaldehyde [1], the rhodium-catalyzed hydrogenation of alkenes as described by Wilkinson [2], the Sharpless epoxidation reaction [3], and more recently palladium-catalyzed cross-coupling reactions for which Suzuki, Heck, and Negishi received the Nobel Prize for Chemistry in 2010 [4]. Traditional homogeneous catalysis has now been extended

Plasmonic Nanomaterials: Characterization and Applications in Organic Synthesis and Catalysis
Edited by Clémence Queffélec and D. Andrew Knight

ISBN 978-981-5129-09-0 (Hardcover), 978-1-003-47406-7 (eBook)
www.jennystanford.com

to include reactions that take place in alternative solvents such as water and ionic liquids [5], enantioselective catalysis [6], and supported molecular catalysis [7], which reflects combining the advantages of homogeneous and heterogenous catalysis. Some of the well-documented advantages of homogeneous catalysis over its heterogeneous counterpart include ease of separation of product from reaction mixture, mild operating conditions, improved selectivities, and amenability toward mechanistic understanding [8]. Although the application of surface plasmon resonance (SPR) to catalytic reactions has been dominated by heterogeneous catalytic systems [9], the exploitation of the localized surface plasmon resonance (LSPR) effect exhibited by certain metal nanoparticles to homogeneous catalysis has started to emerge. Using the design principle in which a molecular homogeneous catalyst, e.g., a ligand-stabilized transition metal complex, is anchored to the surface of a plasmonically active metal nanoparticle, energy can be transferred from the light source, e.g., laser or LED, to the plasmonic support. This energy associated with the light absorption can be utilized for (a) energy transfer sensitization, (b) redox processes of nearby molecule, e.g., catalyst or substrate, and (c) localized heating of the microenvironment, which results in the acceleration of reaction rate (Fig. 4.1).

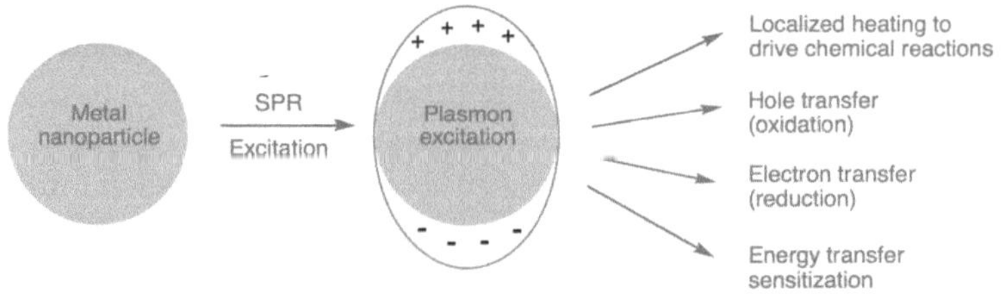

Figure 4.1 Pathways for metal nanoparticle energy coupling to local environment (from Ref. [10]).

Although photochemical homogeneous catalytic reactions have been known for some time [11], the use of LSPR represents a new direction in photo-assisted catalysis. This chapter describes the use of LSPR for the acceleration of homogeneous catalytic reactions based on a recent literature survey with an emphasis on degradation and synthesis.

4.2 Catalytic Hydrolysis

4.2.1 Hydrolysis of Phosphate Esters

The catalytic degradation or cleavage of phosphate and thiophosphate esters, especially those involving a metal-mediated hydrolytic mechanism, is of interest as an approach for the development of materials suitable for the destruction of chemical warfare agents and toxic pesticides as well as selective catalysts for the breakdown of DNA and RNA (artificial endonucleases) [12]. Catalysts based on copper(II) and lanthanides have been some of the most successful. The role of metal ions and metal–ion complexes stabilized with organic ligands such as 2,2′-bipyridine (bpy) and 1,10-phenanthroline in the hydrolytic catalysis of phosphate esters has been known for some time, and the mechanism of these reactions has been extensively studied [13]. The accepted mechanism is that the phosphate ester initially displaces a coordinated water molecule and binds to the metal center. Intramolecular attack by a second coordinated water ligand occurs. Scheme 4.1 shows an example with the $Cu(bpy)^{2+}(aq)$ complex.

These reactions suffer from slow reaction kinetics and high catalyst/substrate ratios and as such are highly amenable to kinetic enhancement using LSPR. In 2012, one of the first homogeneous metal-based catalytic reactions accelerated by SPR was reported by Trammell et al. [15]. The authors described a 10 nm gold nanoparticle (AuNP) platform, which supported a copper(II) bipyridine complex via stable Au-thiol interactions. The supported complex was used for the hydrolysis of methyl parathion, a broad-spectrum thiophosphate insecticide, which is environmentally persistent. Using a low-power (120 mW), 532 nm green laser, it was demonstrated that a rate acceleration of ×2-fold occurred for the hydrolysis of methyl parathion by irradiating the plasmon absorption band of AuNPs, which were capped with the copper-bipyridine catalyst.

The phenomenon of the acceleration of the initial rate of hydrolysis of MeP by the excitation of the plasmon absorption band of the AuNPs could have several mechanistic explanations. For example, energy transfer sensitizing the Cu metal center could create an excited d–d state, which might accelerate the rate of hydrolysis if a dissociative step at the metal center is rate limiting. Another

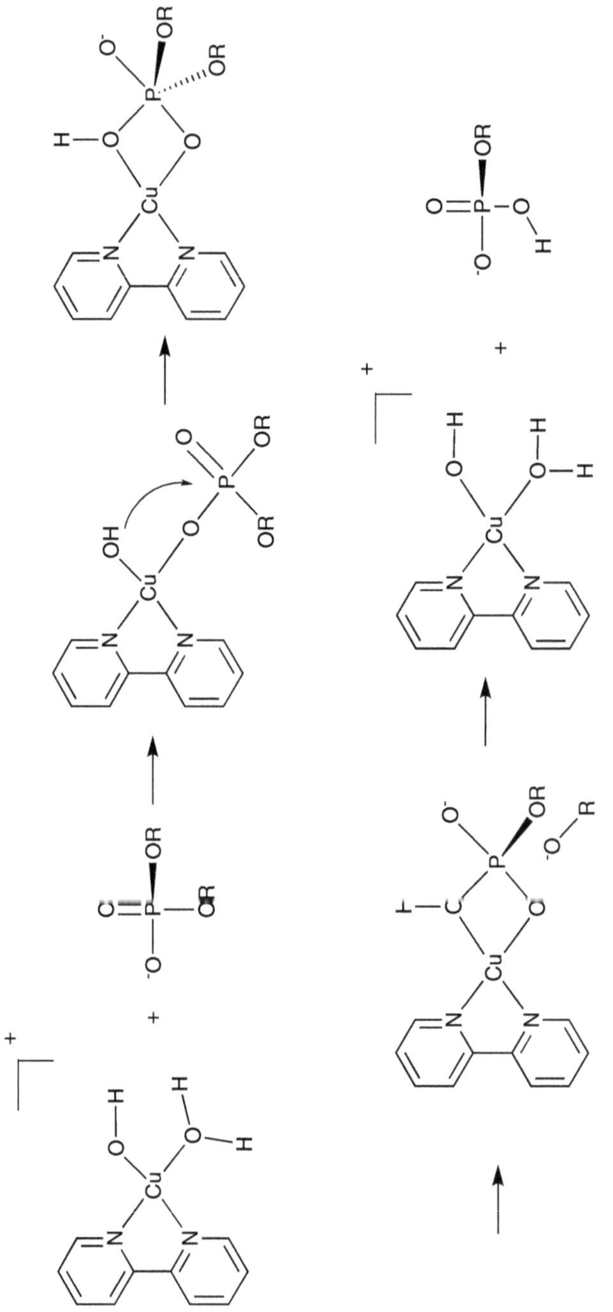

Scheme 4.1 The general mechanism of hydrolytic catalysis of phosphate esters by $Cu(bpy)^{2+}(aq)$ (from Ref. [14]).

explanation maybe the reduction of Cu(II) to Cu(I) by the AuNPs, creating a more active catalyst. However, the simplest explanation is the creation of localized heating of microenvironment near the AuNP surface. Since we see an increase in rate by a factor of ×2, an increase of 10 degrees in the microenvironment would be a common prediction by the Arrhenius equation for this type of reaction. AuNPs are good light-to-heat converters with excitation of their plasmon absorption bands dissipating energy into the microenvironment.

4.3 Catalytic Carbon–Carbon Bond Formation

4.3.1 Ullmann Reaction

Forato et al. [16] developed copper-polypyridine catalysts grafted onto core–shell silver nanomaterial and used this plasmonic active nanocatalyst in catalysis.

They focused on the formation of C–C bonds, and the model reaction studied was the Ullmann reaction, which is a classic example of the catalytic carbon–carbon bond-forming process. This reaction involves the coupling of aryl halides mediated by copper (Scheme 4.2).

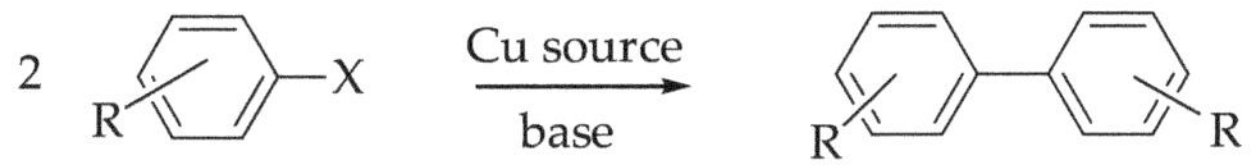

X = I or Br and R = NO_2, OMe ....

Scheme 4.2 Example of the Ullmann reaction catalyzed by copper.

In that regard, the ligand featuring a 2,2′-bipyridine backbone bearing one terminal phosphonic acid tether was obtained in five steps with a 30% overall yield (Scheme 4.3).

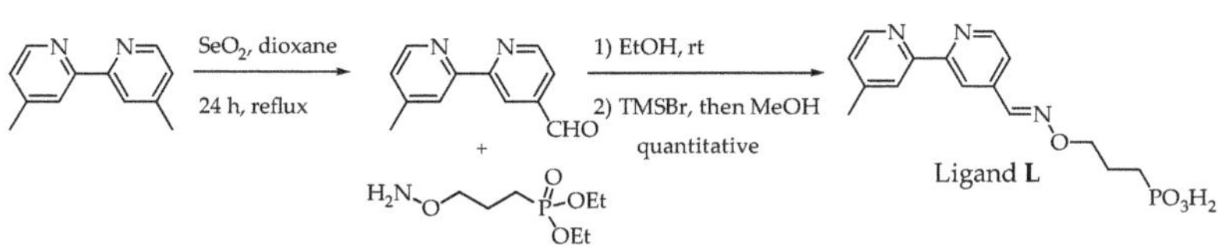

Scheme 4.3 Synthesis of the targeted ligand L.

Core–shell nanoparticles were synthesized using adapted protocols described in the literature [17, 18]. Reduction of silver nitrate with hydrazine monohydrate led to the formation of AgNPs, and the TiO_2 shell was then formed through the hydrolysis of titanium isopropoxide in water, leading to Ag@TiO_2 NPs. From Fig. 4.2a, it can be seen that AgNPs are randomly embedded in a TiO_2 matrix. Then, a water solution of ligand L was added to the Ag@TiO_2 NPs, which led to the grafting of the ligand onto the NPs to form Ag@TiO_2@L NPs (Fig. 4.2b). Finally, an aqueous solution of $Cu(CH_3CN)_4PF_6$ was added to the Ag@TiO_2@L NPs and the final nanocatalyst Ag@TiO_2@L@Cu(I) was obtained (Fig. 4.2c) as spherical nanoparticles with an average diameter of 19 nm.

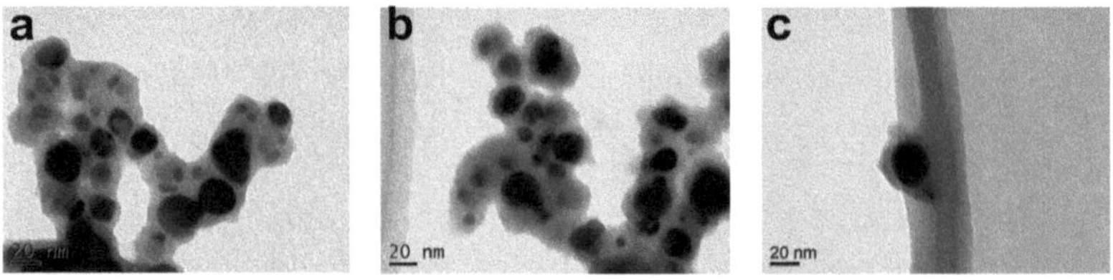

Figure 4.2 TEM images of (a) Ag@TiO_2 NPs, (b) Ag@TiO_2@L NPs, (c) Ag@TiO_2@L-Cu(I) nanoparticles.

UV–vis spectra were recorded to confirm the presence of the plasmonic band that was centered around 445 nm for the nanocatalyst. Besides, the SPR band is quite large, which allows the use of various wavelength for the catalytic reaction irradiation, especially 450 nm or 488 nm wavelengths.

This nanocatalyst was further characterized by surface-enhanced Raman spectroscopy (SERS) [19] and X-ray absorption (XANES and EXAFS) [20]. Thus, the oxidation state of the copper ion was confirmed by XANES and the copper complex geometry grafted on the surface by EXAFS and SERS. The best EXAFS model suggested a low coordinate copper(I) geometry that included one bpy ligand with two nitrogen atoms at 1.90(1) Å and an elongated distance to a second light atom (i.e., C, N, or O) at 2.04(2) Å. Besides, by SERS, different peaks with a small shift in the wavenumber values were attributed to different modes of vibration of bipyridine and confirming that copper was coordinated to the bipyridine skeleton.

Table 4.1 Ullmann coupling reaction under different conditions

Entry	Catalyst	Laser Wavelength	Power	Time	Temperature	Yield
1	CuI (5%) + 2,2′-bipyridine (15%)	No laser	—	24 h	110°C	70%
2	Ag@TiO_2@L@-Cu(I) (≈ 1%)	No laser	—	24 h	110°C	<5%
3	Ag@TiO_2	488 nm	220 mW	8 h	T.A.	No conversion
4	Ag@TiO_2@- L@Cu(I) (≈ 1%)	488 nm	220 mW	5 h	T.A.	55%
5	Ag@TiO_2@L@-Cu(I) (≈ 1%)	488 nm	220 mW	8 h	T.A.	65%

Then, as a proof of concept, the simplest example of the Ullmann reaction was studied using iodobenzene reagent, which leads to the formation of a biphenyl product. The reaction was carried out under conventional conditions, inspired by protocols described in the literature, and the final product was obtained in 70% yield after 24 h (Table 4.1, entry 1). Then this reaction was performed using a colloidal solution of copper(I) NPs (copper/substrate = 5 mol%). In the absence of laser irradiation or in the absence of a catalyst but with laser irradiation, no conversion was observed (Table 4.1, entries 2 and 3). Using a continuous argon laser at λ_{exc} = 488 nm with power equal to 220 mW resulted in the desired product in 55% yield after 5 h and 65% yield after 8 h (Table 4.1, entries 4 and 5).

This was the first example published proving that molecular plasmonics could be used to enhance a catalytic reaction.

4.3.2 Henry's Reaction

The same group studied AuNPs onto which they grafted a copper(II) complex that was active in Henry's reaction under green LED irradiation [21].

Henry's reaction is a base-catalyzed C–C bond-forming reaction between nitroalkanes and aldehydes or ketones, which leads to β-nitroalcohol products (Scheme 4.4). Besides, tremendous work has been done on the development of the asymmetric version, and most examples of the asymmetric Henry reaction involve the use of chiral copper complexes [22–26].

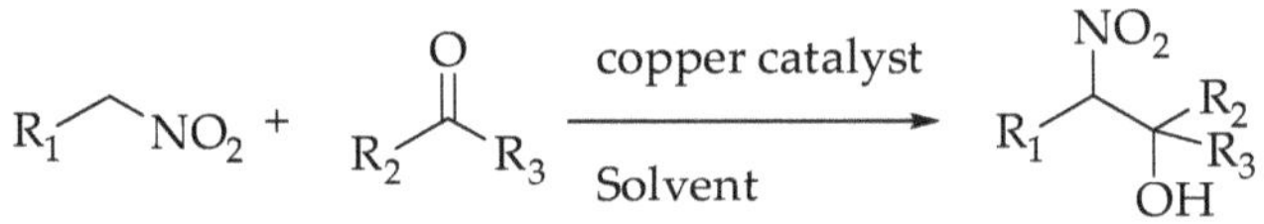

Scheme 4.4 Example of Henry's reaction catalyzed by copper.

Regarding the ligand, it possessed a bipyridine backbone and a tether having a terminal dithiol unit that ensures the anchoring of the ligand onto the surface of AuNPs [27]. The synthesis is described in Scheme 4.5, and the ligand is obtained with an overall yield of 15.5%.

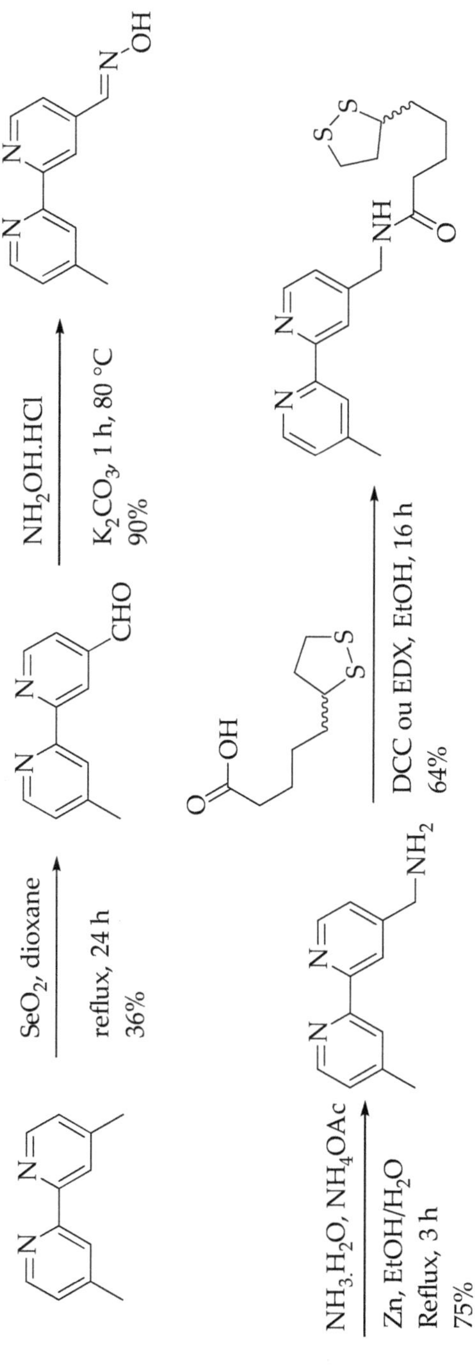

Scheme 4.5 Synthesis of the ligand used in Henry's reaction.

Then, to obtain the copper(II) complex, the ligand was dissolved in degassed ethanol and copper acetate was added. The solution was stirred for 2 days, and the copper(II) complex was obtained.

The AuNPs were synthesized through a "reverse Turkevich" protocol [28], and spherical AuNPs were obtained with an average size of 20–30 nm. The colloids were then transferred into a DMF solvent, and the copper(II) complex was added to this solution, leading to the grafting of the complex onto the AuNP surface. The anchoring was demonstrated by SERS where the breaking of the dithiol S–S bond was observed near 509 cm^{-1}. An Au/Cu ratio of ca. 350 was measured by ICP-AES.

To perform Henry's reaction, 4-nitrobenzaldehyde and nitromethane were used as starting materials in DMF and the nanocatalyst (Ncat) was added and the reaction was irradiated with green light (of wavelength 525 nm). The formation of nitroaldol was assessed by ^1H NMR (a reaction conversion was estimated) and SERS, and the results were compared to dark conditions. After 30 min, the conversion under irradiation reached 82%, while under dark, the conversion was around 56% (Fig. 4.3).

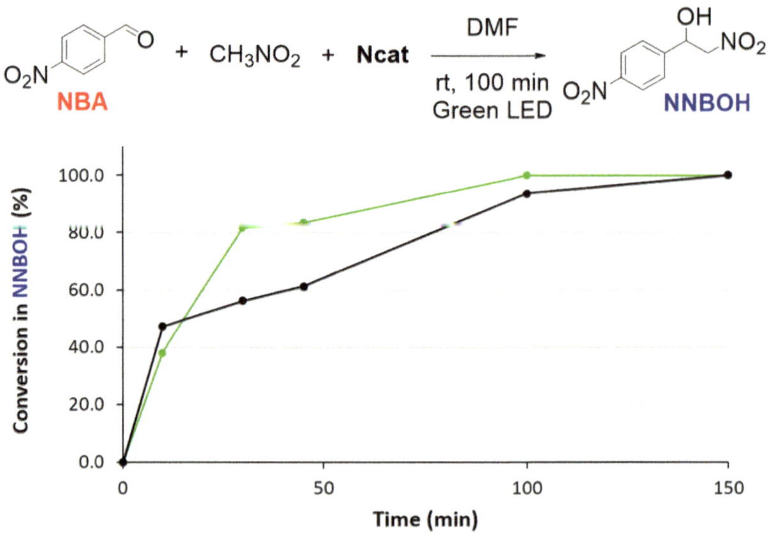

Figure 4.3 Conversion over time for Henry's reaction performed with Ncat in DMF under dark conditions (black curve) and green LED irradiation (green curve). Reproduced from Ref. [21].

Investigations and specific treatments of the Raman signals, for example the principal component analysis (PCA), allowed to observe the decrease in the Raman signal at 1710 cm^{-1} characteristic of the aldehyde group in favor of the appearance of the signal at 1355 cm^{-1} assigned to the nitro group on the aldol, suggesting again the formation of the final product. The difference was more important when the reaction was irradiated with green light.

Those results showed that the rate of Henry's reaction was enhanced by the interaction of the Cu(II) complex grafted onto Au colloidal NPs under green light adsorption, and it was proven that the reaction was much faster with the nanocatalyst under green LED irradiation than in the classic conditions (ca. 100% conversion after 100 min versus ≈ 50% conversion after 16 h).

References

1. Keith, J. A. and Henry, P. M. The mechanism of the Wacker reaction: A tale of two hydroxypalladations. *Angew. Chem. Int. Ed.* 2009, **48**(48), 9038–9049. https://doi.org/10.1002/anie.200902194.
2. Osborn, J., Jardine, F., Young, J., and Wilkinso, G. Preparation and properties of tris(triphenylphosphine)halogenorhodium(1) and some reactions thereof including catalytic homogeneous hydrogenation. *J. Chem. Soc. Inorg. Phys. Theor.* 1966, 12, 1711. https://doi.org/10.1039/j19660001711.
3. Katsuki, T. and Sharpless, K. B. The first practical method for asymmetric epoxidation. *J. Am. Chem. Soc.* 1980, **102**(18), 5974–5976. https://doi.org/10.1021/ja00538a077.
4. The Nobel Prize in Chemistry 2010. NobelPrize.org. https://www.nobelprize.org/prizes/chemistry/2010/summary/ (accessed 2023-01-31).
5. Giernoth, R. Homogeneous catalysis in ionic liquids. In *In Situ NMR Methods in Catalysis*: *Topics in Current Chemistry*, Bargon, J. and Kuhn, L. T. (Eds.). Springer: Berlin, Heidelberg, 2007, pp. 1–23. https://doi.org/10.1007/128_016.
6. McMorn, P. and Hutchings, G. J. Heterogeneous enantioselective catalysts: Strategies for the immobilisation of homogeneous catalysts. *Chem. Soc. Rev.* 2004, **33**(2), 108–122. https://doi.org/10.1039/B200387M.

7. Fan, J. and Gao, Y. Nanoparticle-supported catalysts and catalytic reactions: A mini-review. *J. Exp. Nanosci.* 2006, **1**(4), 457–475. https://doi.org/10.1080/17458080601067708.
8. Thomas, S. J. M. The advantages of exploring the interface between heterogeneous and homogeneous catalysis. *ChemCatChem* 2010, **2**(2), 127–132. https://doi.org/10.1002/cctc.200900275.
9. Adleman, J. R., Boyd, D. A., Goodwin, D. G., and Psaltis, D. Heterogenous catalysis mediated by plasmon heating. *Nano Lett.* 2009, **9**(12), 4417–4423. https://doi.org/10.1021/nl902711n.
10. Knight, D. A., Nita, R., Moore, M., Zabetakis, D., Khandelwal, M., Martin, B. D., Fontana, J., Goldberg, E., Funk, A. R., Chang, E. L., and Trammell, S. A. Surface plasmon resonance promotion of homogeneous catalysis using a gold nanoparticle platform. *J. Nanoparticle Res.* 2014, **16**(6), 2400. https://doi.org/10.1007/s11051-014-2400-8.
11. Hoffmann, N. Homogeneous photocatalysis with organometallic compounds. In *Applied Homogeneous Catalysis with Organometallic Compounds*, John Wiley & Sons Ltd., 2017, pp. 1097–1134. https://doi.org/10.1002/9783527651733.ch17.
12. Liu, C., Wang, M., Zhang, T., and Sun, H. DNA hydrolysis promoted by di- and multi-nuclear metal complexes. *Coord. Chem. Rev.* 2004, **248**(1), 147–168. https://doi.org/10.1016/j.cct.2003.11.002.
13. Morrow, J. R. and Trogler, W. C. Hydrolysis of phosphate diesters with copper(II) catalysts. *Inorg. Chem.* 1988, **27**(19), 3387–3394. https://doi.org/10.1021/ic00292a025.
14. Moore, M., Knight, D. A., Zabetakis, D., Deschamps, J. R., Dressick, W. J., Chang, E. L., Lascano, B., Nita, R., and Trammell, S. A. Electronic effects on the reactivity of copper mono-bipyridine complexes. *Inorganica Chim Acta* 2012, **388**, 168–174. https://doi.org/10.1016/j.ica.2012.03.020.
15. Trammell, S. A., Nita, R., Moore, M., Zabetakis, D., Chang, E., and Knight, D. A. Accelerating the initial rate of hydrolysis of methyl parathion with laser excitation using monolayer protected 10 nm Au nanoparticles capped with a Cu(Bpy) catalyst. *Chem. Commun.* 2012, **48**(34), 4121–4123. https://doi.org/10.1039/c2cc30850a.
16. Forato, F., Talebzadeh, S., Bujoli, B., Queffelec, C., Trammell, S. A., and Knight, D. A. Core–shell Ag@TiO_2 nanocomposites for low-power blue laser enhanced copper(I) catalyzed Ullmann coupling. *Chemistryselect* 2017, **2**(2), 769–773. https://doi.org/10.1002/slct.201601788.
17. Sakai, H., Kanda, T., Shibata, H., Ohkubo, T., and Abe, M. Preparation of highly dispersed core/shell-type titania nanocapsules containing a

single Ag nanoparticle. *J. Am. Chem. Soc.* 2006, **128**(15), 4944–4945. https://doi.org/10.1021/ja058083c.

18. Angkaew, S. and Limsuwan, P. Preparation of silver-titanium dioxide core–shell (Ag@TiO_2) nanoparticles: Effect of Ti-Ag mole ratio. *Procedia Engineering* 2012, **32**, 649–655.

19. Forato, F., Talebzadeh, S., Rousseau, N., Mevellec, J.-Y., Bujoli, B., Knight, D. A., Queffelec, C., and Humbert, B. Functionalized core–shell Ag@ TiO_2 nanoparticles for enhanced Raman spectroscopy: A sensitive detection method for Cu(Ii) ions. *Phys. Chem. Chem. Phys.* 2019, **21**(6), 3066–3072. https://doi.org/10.1039/c8cp07504b.

20. Queffélec, C., Forato, F., Bujoli, B., Knight, D. A., Fonda, E., and Humbert, B. Investigation of copper oxidation states in plasmonic nanomaterials by XAS and Raman spectroscopy. *Phys. Chem. Chem. Phys.* 2020, **22**(4), 2193–2199. https://doi.org/10.1039/C9CP06478H.

21. Gimeno, L., Picardi, G., Planchat, A., Knight, D. A., de la Chapelle, M. L., Humbert, B., and Queffelec, C. Improving the rate of the copper-catalyzed Henry reaction by surface plasmon excitation of gold nanoparticles. *Catal. Sci. Technol.* 2021. https://doi.org/10.1039/D1CY01788H.

22. Zhang, L., Wu, H., Yang, Z., Xu, X., Zhao, H., Huang, Y., and Wang, Y. Synthesis and computation of diastereomeric phenanthroline-quinine ligands and their application in asymmetric Henry reaction. *Tetrahedron* 2013, **69**(49), 10644–10652. https://doi.org/10.1016/j.tet.2013.10.010.

23. Didier, D., Magnier-Bouvier, C., and Schulz, E. Charge-transfer interactions: An efficient tool for recycling bis(oxazoline)-copper complexes in asymmetric Henry reactions. *Adv. Synth. Catal.* 2011, **353**(7), 1087–1095. https://doi.org/10.1002/adsc.201000934.

24. Cho, J., Chun, M. K., Nayab, S., and Jeong, J. H. Synthesis and structures of copper complexes bearing unsymmetric derivatives of (R,R)-1,2-diaminocyclohexane: An efficient catalyst for asymmetric Henry reaction. *Appl. Organomet. Chem.* 2019, **33**(7), e4955. https://doi.org/10.1002/aoc.4955.

25. Gao, E., Li, M., Duan, L., Li, L., and Li, Y.-M. Isosteric expansion of the structural diversity of chiral ligands: Design and application of proline-based N,N′-dioxide ligands for copper-catalyzed enantioselective Henry reactions. *Tetrahedron* 2019, **75**(37), 130492. https://doi.org/10.1016/j.tet.2019.130492.

26. Rossetti, A., Landoni, S., Meneghetti, F., Castellano, C., Mori, M., Dugoni, G. C., and Sacchetti, A. Application of chiral bi- and tetra-dentate bispidine-derived ligands in the copper(II)-catalyzed asymmetric Henry reaction. *New J. Chem.* 2018, **42**(14), 12072–12081. https://doi.org/10.1039/C8NJ01930D.
27. Picardi, G., Humbert, B., de la Chapelle, M. L., and Queffelec, C. Surface modification of Au nanoparticles with heteroleptic Cu(I) diimine complexes. *J. Phys. Chem. C* 2020, **124**(22), 11902–11912. https://doi.org/10.1021/acs.jpcc.0c01881.
28. Sivaraman, S. K., Kumar, S., and Santhanam, V. Monodisperse sub-10 nm gold nanoparticles by reversing the order of addition in Turkevich method – The role of chloroauric acid. *J. Colloid Interface Sci.* 2011, **361**(2), 543–547. https://doi.org/10.1016/j.jcis.2011.06.015.

Chapter 5

SERS and TERS and Their Applications in Organic Synthesis or Catalysis

Gennaro Picardi,[a,b] Mathieu Edely,[a] and Marc Lamy de la Chapelle[a]
[a]*Le Mans University, Laboratoire IMMM, UMR6283, Avenue Olivier Messiaen 72085 Le Mans Cedex 9, France*
[b]*Nantes University, CNRS, CEISAM, UMR 6230, F-44000 Nantes, France*
marc.lamydelachapelle@univ-lemans.fr

5.1 Surface-Enhanced Raman Spectroscopy

The most limiting aspect of Raman spectroscopy as a generally applicable analytical technique is the extremely low cross section ($d\sigma/d\Omega$ = 10^{-31} to 10^{-28} $cm^{-2}sr^{-1}$) of the Raman scattering process compared with the much higher probability of other optical processes, for example fluorescence, which already meets the sensitivity requirements for single-molecule detection (Fig. 5.1). Thus, powerful exciting sources, sensitive detectors, and a large number of scattering molecules are normally needed to record a Raman spectrum. The last constraint would make Raman

Plasmonic Nanomaterials: Characterization and Applications in Organic Synthesis and Catalysis
Edited by Clémence Queffélec and D. Andrew Knight

ISBN 978-981-5129-09-0 (Hardcover), 978-1-003-47406-7 (eBook)
www.jennystanford.com

spectroscopy rather uninformative for catalytic studies at surfaces, usually dealing with a small number of adsorbed molecules, down to the (sub) monolayer level.

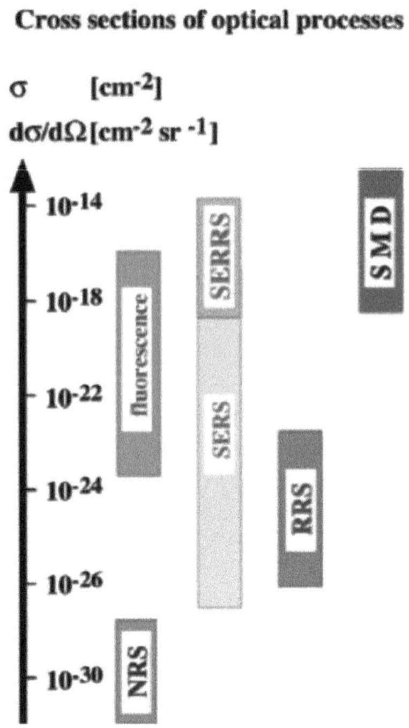

Figure 5.1 Cross sections of optical processes. NRS = normal Raman scattering; RRS = resonance Raman scattering; SERS = surface-enhanced Raman scattering; SERRS = surface-enhanced resonance Raman scattering; SMD = single-molecule detection. Adapted from Ref. [19].

In contrast to this expectation, a strongly enhanced Raman signal can be observed from a large number of chemically different molecules when adsorbed at a metal surface (Ag, Au, and Cu deliver the strongest signal enhancement). A roughening or, more precisely, nano-structuring of the metal surface is the key factor for the observation of the enhanced signal. This effect is known as surface-enhanced Raman scattering, and the related surface-enhanced Raman spectroscopy (SERS) boosted the Raman efficiency to a level comparable or even superior to fluorescence spectroscopy. With respect to the chemical identification of an unknown analyte, Raman spectroscopy is more revealing than fluorescence, since it interrogates molecular vibrations and yields a number of sharp and characteristic bands as compared to the very broad and often unspecific fluorescence bands. Moreover, fluorescence is effectively limited only to a class of large organic molecules (fluorescent dyes).

The first experimental observations of the SERS effect were made on roughened metal surface used as electrodes in spectroelectrochemical studies. In an attempt to increase the surface effective area, and thus the number of adsorption sites available to pyridine molecules on an Ag electrode, Fleischmann and Hendra [1] recorded in 1974 unexpectedly high intensity of Raman-scattered light (corresponding to the molecular vibration of pyridine) by applying cyclic potential steps to the metal electrode. A few years later, van Duyne et al. [2] and independently Albrecht and Creighton [3] repeated these measurements, presenting a different and correct, although astonishing, interpretation. The Raman cross section of molecules receives a huge enhancement upon adsorption on the roughed metal surfaces.

It is commonly accepted that the SERS effect arises due to two different and concomitant mechanisms: electromagnetic effect and chemical effect. Together, they may provide enhancement factors between 10^4 and 10^{10} [20–22].

5.1.1 Electromagnetic Effect

The electromagnetic effect involves the optical excitation of localized surface plasmon (LSP) resonances in the substrate metal and the resulting amplified electromagnetic field locally present at the surface. Chapter 1 introduced surface plasmon excitation at the (theoretically infinite) flat interface between a metal and a dielectric and the case of localized surface plasmon modes, where the metal surface cannot be considered planar with respect to the wavelength of the exciting radiation (surface curvature much smaller than the wavelength of visible light). In both cases, the excitation of surface plasmon modes generates evanescent electromagnetic waves, which can be delocalized, freely traveling over the whole (supposedly unbounded) planar surface, but still evanescently decaying in the direction of the surface normal or localized at the (finite, bounded) surface of a nanostructure. In particular, for spherical metal nanoparticles, exact analytical solutions to the Maxwell equations exist. Mie solutions reveal the wavelength positions where the absorption and scattering of the incident light is maximum due to the resonant excitation of the (dipolar) plasmon resonances, describing the induced, coherent oscillations of the electrons at the surface. From the Mie solutions, it is also possible to evaluate the amplified intensity of the electromagnetic field near the metal surface with respect to the incident field.

The presence of a locally amplified electromagnetic field at or very near a metal surface has a strong effect on the intensity of the Raman signal from molecules located (adsorbed) at or very near the metal surface. The intensity of the Raman signal at a given frequency is (approximately) proportional to the fourth power of the electric field intensity at the molecule location. Thus, the molecules that experience an amplified field give an enhanced Raman signal (Fig. 5.2). Due to the fourth power relationship, even a moderate field amplification may result in a very strong enhancement of the Raman signal. At certain hot spots, the field can be amplified up to 30 times and the corresponding Raman signal is 10^6 times enhanced!

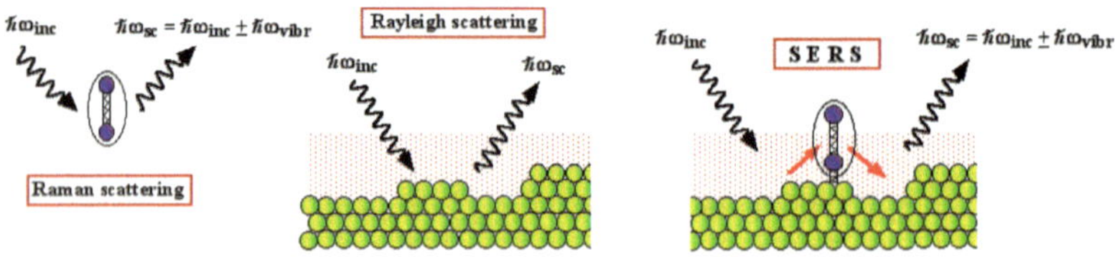

Figure 5.2 Graphical representation of different optical processes. (Left) Raman scattering generically considering non-interacting molecules in the liquid or gaseous state. The incident photon has frequency ν_{inc}, and the Raman-scattered photon has a different frequency ν_{sc}. The difference in photon energy corresponds to a quantum of vibrational energy $h\nu_{vibr}$. (Center) The elastic (Rayleigh) scattering of propagating (far-field) light by a rough metal substrate under continuous monochromatic illumination. LSP modes are excited at the resonant frequency ν_{inc}; these surface resonances after a very short time decay radiatively emitting a photon ν_{sc}. In this case, the incident and scattered photons have the same energy. An enhanced electromagnetic field (red dotted area) is present very close to the surface (in the nearfield), resulting from the excitation of the LSP resonances. (Right) Surface-enhanced Raman scattering for the same molecule now interacting with the rough surface. The molecule is embedded in the enhanced electromagnetic field, and the Raman signal is amplified. Comparison with Fig. 5.3 helps recognizing the different processes involved.

Figure 5.3 gives a diagrammatic representation of the SERS process taking place at a roughened metal surface (a nonplanar, nanostructured surface, or else a generic nanoparticle that presents LSP resonances) due to illumination with a monochromatic light source. Consider first the illuminated surface with no adsorbed molecules (on the left). LSP modes are excited by the absorption of a photon with energy $h\nu_{inc}$; this frequency is supposed to be in resonance, i.e., to fall within the LSP extinction curve of the nanoparticle(s). The metal particle is raised to an excited state and will relax to the ground state (relaxation time in the order of 10–15 s) with the annihilation of the LSP mode and the creation of a second photon. This process is a case of Rayleigh scattering, with the nanoparticle acting as the light scatterer. The frequency of the emitted photon remains the same as for the incident photon, $\nu_{sc} = \nu_{inc}$, but the other components of the wave vector (direction, phase) may have changed, $K_{sc} \neq K_{inc}$.

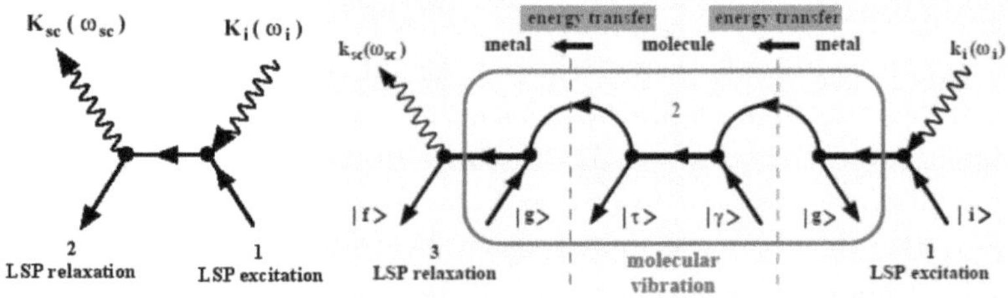

Figure 5.3 (Left) Rayleigh scattering of light by LSP excitation from a metal nanoparticle. (Right) Raman scattering enhanced by a metal nanoparticle (which is now considered to be covered by a layer of molecules). The total energy states of the hybrid system are approximated as the product of the uncorrelated states of the molecule and the metal: |i> = |g> | γ>, |f> = |g> |τ>. | γ> and |τ> are different vibrational states of the molecule; |g> is the ground electronic states of the nanoparticle. In step 1, the nanoparticle is excited from |g> to a state |h> and then de-excited back to |g>, transferring energy to the molecule. In step 3, the nanoparticle is excited from |g> to a state |h′> by receiving energy from the molecule and then de-excited back to |g> by emitting the scattering photon. The plasmon excited states |h> and |h′> differ by the vibrational quantum of energy deposited in the molecule (going from | γ> to | τ>), which is also the energy difference between the incident and scattered photons. Adapted from Ref. [23].

A different optical process arises if the metal surface is covered by adsorbate molecules: The energy stored in the metal as LSP may be transferred via dipole–dipole interaction to the molecule. Vibrational modes may be activated before a further energy transfer occurs, this time from the molecule back to the metal. Now the scattered photon has a different energy, and the difference in energy is the quantity gained or lost by the molecule through its vibration. This is an inelastic scattering or Raman process (involving both the molecule and the metal surface). As reported in the diagram, the whole process may be figuratively divided into three resonant steps: Most importantly, the LSP resonances take part twice (but with slightly different energies, $h\nu_{inc}$ and $h\nu_{sc}$); they are twice excited and twice de-excited. Compared with the normal resonance Raman process, where the molecule interacts directly with the incident electromagnetic field, the field experienced by the adsorbed molecule is the local field enhanced by the induced oscillations of the electrons in the metal particle.

All optical processes may profit from this local field enhancement, and surface-enhanced infrared spectroscopy (SEIRA) and metal-enhanced fluorescence (MEF) exist as well. These are single-photon absorption processes, and the signal enhancement is "only" the square of the field amplification, since there is only one photon–LSP interaction. The substantially greater popularity of SERS lies in the fourth power relationship stemming from the LSP excitation taking part twice in the overall optical scattering process: First, the incident field (photon) interacts with the LSP with an energy $h\nu_{inc}$, and then the scattered field at the Raman-shifted frequency $h\nu_{sc}$ interacts with the LSP (Figs. 5.2 and 5.3). Indeed, the small difference in energy between $h\nu_{inc}$ and $h\nu_{sc}$ (a molecular vibrational quantum compared to the electronic energies of the LSP resonances) is the reason why the SERS fourth power law is an approximation. More precisely one should multiply the amplification of the field at the energy $h\nu_{inc}$ (the square of this field intensity) by the amplification of the field at the energy $h\nu_{sc}$ (the square of this different field intensity). This is approximated by considering the same amplification at the two close energies and for the Raman signal enhancement $|E_{inc}|^{2*}|E_{sc}|^2 \sim |E_{inc}|^4$.

Hot spots are particular locations on the surface where the electromagnetic field is strongly enhanced due to the geometry of the corresponding feature (remember that the intensity of the LSP resonances depends on the size, shape, and metal dielectric constant; see Chapter 1 for more details). A simple example is the case of elongated nanoparticles (nanorods) illuminated with light polarized along the long axis. In this case, the strongest electron oscillations and the strongest field intensity are localized near the apexes, and the molecules placed near the apexes will provide a much stronger SERS signal than the molecules placed near the sides of the rod. The concept of hot spots is reminiscent of catalytic active sites: On a catalytic surface, a few sites are mostly responsible for the catalytic activity, the rest of the surface being rather inactive.

Similarly, the molecules adsorbed at the hot spots are responsible for most of the SERS signal, while all the other molecules, despite being illuminated by the laser, provide only a normal, unenhanced Raman signal, almost negligible with respect to the SERS. Thus, the SERS signal originates from a much smaller number of molecules than the number that can be estimated by taking in account the laser spot size and the illuminated geometrical area. Furthermore, the enhanced electromagnetic field expands only up to ~100 nm away from the surface, meaning that the volume sensed by SERS is very limited. There is a consensus on the magnitude and spatial decay of the enhanced electromagnetic field intensity with the distance from the nanostructure: The strongest enhanced fields decay the fastest (within 10 nm); the weaker fields, generated by larger structures, decay more slowly. This is very relevant for SERS, since an exceedingly small number of molecules located next to a very (and small) hot spot contribute to the vast majority of the SERS signal; the vast majority of molecules located in "colder" regions do not contribute much. This is the tenet for the very high SERS sensitivity and its potential for single-molecule detection.

Concerning the shape and size of these hot spots, other than sharp protrusions (with end radius <30 nm), even stronger fields are present at the nanogaps or nano-constrictions formed between two almost touching metal structures for the incident light polarized across the gap. The representative example is a dimer of spherical (colloidal) nanoparticles approaching to within ~2 nm or, in the case of lithographically fabricated structures, the bowtie antenna geometry. In both cases, the field within the nanogap results from the nearfield overlap and interaction of the fields individually supported by each constituent nanoparticle, a sort of constructive interference with multiplicative effects. Figuratively speaking, approaching a second plasmonic object has the consequence to further localize, "compress" the LSP excitation of the other one. As before, the general rule of thumb is as follows: The more localized the excitation, the strongest the resulting field.

SERS is inherently a nearfield probe technique. The sensing electromagnetic field interacting with the molecules responsible for the SERS signal is not the field directly associated with the propagating (far-field) incident light, but it is the enhanced field present very close to particular metal nanostructures and associated with the LSP excitation characterized by nearfield optical behavior (non-propagating, evanescence). The excited electromagnetic field is no longer brought to the sample by macroscopic optical elements, such as lenses or microscope objectives. Instead, the excited field is delivered by nano-objects, such as colloids or supported nanostructured (SERS), tapered fiber tips (aperture SNOM) or sharp metal tips (aperture-less SNOM and TERS).

5.1.2 Chemical Effect

The chemical enhancement mechanism involves a photo-assisted charge transfer (CT) between the metal and the adsorbate at specific sites of the surface. The adsorbed molecules may specifically interact with the metal, creating a new set of energy states that are available for the Raman process. Indeed, the best way to view the chemical enhancement is as an analog of the resonance Raman scattering process associated with these new CT states denoting the metal/molecule hybrid. The isolated molecule does not possess these additional states, and there is no chemical enhancement (of course, the molecule, e.g., a dye, can still be "in resonance" with the exciting wavelength). The chemical enhancement is a resonant amplification of the molecular Raman cross section brought by the metal surface or rather by the interaction with the metal surface. In a more physical picture, the proximity of the metal surface modifies the molecular polarizability and hence the Raman cross section.

Consider the energy level diagram in Fig. 5.4, with the generic band structure of the metal on the left and the molecular energy states on the right. In the first step (1), the incident radiation $h\nu_i$ excites an electron within the metal conduction band to an empty state in the half-filled band. This charge is transferred from the metal to the molecule (2). E_{CT} is the electronic state given by the mixing of a molecule non-bonding state and an empty metal state of appropriate energy. If the molecule is left in a different vibrational state (3), after a back-charge transfer from molecule to metal (4), the e^-h^+ pair and, thus, the photon leaving our system (5) have a (slightly) different energy. The final step (5) is the radiative annihilation of the e^-h^+ pair. As before, the whole process (1 to 5) can be, in the whole, considered a photon-scattering process, mediated in this case by the excitation of an electron–hole pair, while for the electromagnetic enhancement, it was discussed in terms of LSP scattering (Fig. 5.2).

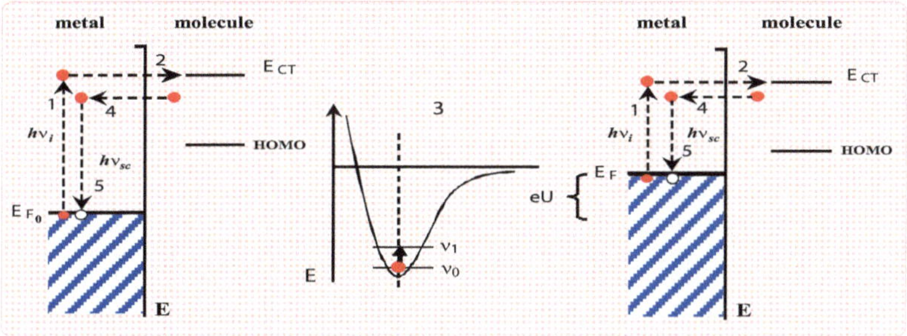

Figure 5.4 Energy diagram for a charge transfer (CT) process at the metal/adsorbed molecule interface. In the center, the Morse potential for the charge transfer state is schematically depicted to denote the molecular Raman scattering (step 3). On the right and left, two different positions for the Fermi level are shown, as modified by an applied potential U. This shows that in the two cases, different $h\nu_i$ will be required to have the chemical enhancement acting. For a given E_{CT} (and Fermi level of the metal), certain exciting wavelengths may be better in resonance than others and the chemical enhancement may be more or less effective in enhancing the Raman signal.

If the metal acts as an electrode in an electrochemical environment, the position of the Fermi level E_F (relative to the zero energy at infinite) may be shifted at will by changing the applied voltage U. The optimal excitation energy for the overall resonance process is thus defined by the following:

$$\hbar\nu_i = E_{CT} - [E_F(0) + eU] \qquad (5.1)$$

This dependency has been confirmed by the observation of different (potential dependent) Raman intensity maxima for different excitation laser frequencies [24, 25].

The latest work of Otto interrogated if molecules adsorbed at certain surface structures were likely to contribute more to the chemical enhancement [26]. Similar to the electromagnetic enhancement in which the LSP excitation with the associated electromagnetic field strongly concentrates only at certain hot spots, the chemical enhancement may be unequally operating for all the adsorbed surfaces and there might be hot spots of chemical enhancement. With respect to the ideal atomically smooth terrace, surface "defective" sites support different adsorption geometry, and also the metal DOS may locally vary, implying that the $e^{-}h^{+}$ pair excitation and charge transfer process involve modified electronic states of the adsorbed molecule and of the metal. The chemical enhancement requires atomic-scale roughness, a length scale at least 10 times smaller than that of the nanostructures supporting LSP resonances in the visible range. The hot spots for the electromagnetic enhancement may spatially overlap (or not) the hot spots of chemical enhancement. If the two mechanisms act cooperatively, the inelastic scattering of the LSP takes place via the charge transfer process. The coexistence and/or cooperation of the two mechanisms complicate the estimation of the amount of Raman signal enhancement, which can be assigned as related to the "pure" LSP excitation and to the CT process. Very conservatively, the electromagnetic enhancement is considered to contribute the most (between 10^3 and 10^8), and the chemical enhancement further contributes with one to three further orders of magnitude, depending on the molecular structure. This leads to a further comment: The Raman signal intensity depends on the intensity of the local electromagnetic field. The electromagnetic enhancement can be seen as an "external" mechanism, external and "independent" from the characteristic of the molecule(s) (orbital energies) on which it is acting. For this reason, the electromagnetic mechanism is supposed to enhance the Raman signal of (chemically) different molecules by the same amount, as long as the compared molecules occupy exactly the same location on the surface and have similar size. On the contrary, the mechanism of the chemical enhancement is "internal" to the molecule, in the sense that it very much depends on the molecular structure and its interaction with the metal. A given molecule may profit from the chemical enhancement at a specific metal surface site, but another molecular structure adsorbed exactly at the same location may not. Note anyhow that this net separation of electromagnetic and chemical enhancement underrates the importance of cooperating mechanisms (at the same physical location) with multiplicative effects on the intensity of the Raman signal.

Due to their different nature, the electromagnetic enhancement is a long-range effect (a few nanometers from the metal surface), while the chemical enhancement is only active at very short ranges and can be described as the first layer effect. This should be clear since the chemical enhancement requires physical overlapping of the molecular orbitals with the metal surface electronic states to form the charge transfer hybrid state. Molecules not in contact with the metal surface cannot profit from this mechanism. On the other hand, the electromagnetic enhancement decreases more slowly with increasing distance from the metal surface. Conservatively, it may extend for up to 10 nm; thus, it may act, although less efficiently, also on molecules located at a distance of a few nanometers from the surface.

SERS represents an excellent platform to explore the physics of LSP, and since the first fortuitous discovery, it has grown to a mature technique, valued by surface scientists, which has furthermore found field applications outside research laboratories. This is due largely to the technical advancements in fabricating in a reproducible way efficient SERS substrates and also to the good understanding of the chemical, physical picture of the mechanisms involved. Still, some finer aspects, dealing with the chemical enhancement and the role of atomic-scale roughness, or the electromagnetic field distribution, propagation, and field gradients at that same scale are still subject of experimental and theoretical investigations beyond the semiclassical treatment. The high sensitivity and chemically detailed fingerprint response of SERS can be exploited in sensors technology. SERS-based sensors are being developed, in parallel with more established LSP resonance shift technology, for the ultrasensitive detection of environmental pollutants, explosives, drugs, and various biomolecules [27, 28].

Manifestly, SERS also presents many limitations, which were very soon recognized. First, SERS is effective only for Ag, Au, and Cu. The DOS of these metals and the energy values for their interband transitions are such that they mostly resemble free electron metals in the visible range. Large negative values of the real part of the wavelength-dependent dielectric constant and small values of the imaginary part, associated to plasmonic losses, are prerequisites for intense and sharp plasmon resonances. More specifically, the strongest electromagnetic enhancement and SERS effect are displayed by Ag, followed by Au and Cu. Au has the valuable advantage over Ag to be very stable toward oxidation in air; the SERS activity of Ag rapidly decays over a few hours from exposing the surface.

Different strategies have been devised to study by SERS molecular adsorption also at other metal surfaces. One example is the coating of a SERS-active substrate with a few layers of a plasmonically inactive metal (Pt or Pd), essentially borrowing the SERS activity from the buried metal. While those efforts enabled observation of SERS signal from (sub-) monolayers on many surfaces other than the three coinage metals, the main asset of SERS is compromised: The overall signal is moderate at best. Then, even for Ag, Au, and Cu, another limiting requisite is the necessity to have surface roughness or, more generally, a nanostructured surface. This proscribes SERS studies at atomically flat and well-oriented surfaces, which are the archetypical surfaces to understand and optimize surface catalytic activity. Tip-enhanced Raman spectroscopy cleverly bypasses this and other constraints by creating an external hot spot. Indeed, SERS requires hot spots; in many cases, these are randomly generated, e.g., small aggregates of colloidal nanoparticles, nanogaps in roughened surfaces. This randomness limits the usefulness of SERS as a quantitative analytical technique, meaning that the exact number of probed molecules mostly depends on the number and spatial extension of the hot spots; usually simplifying assumptions are made.

Ideally, the shape, number, and distribution of hot spots should be controlled. Fabrication techniques capable to very precisely "engineer" hot spots exist (lithographies and more elaborate nano-structuring techniques), but these are not cheap and/or not intended for mass production. The nanofabrication of efficient and identically enhancing SERS substrate to be used, for example, as diagnostic biosensors in point-of-care facilities is a challenge currently tackled.

From an experimental point of view, the instrumentation required for SERS is the same as for normal Raman spectroscopy. The confocal microscope configuration is perfectly suited since it focusses the light to a spot area of a few micron squares where, depending on the SERS substrates, only a few or a large number of hot spots are simultaneously illuminated. Moreover, the microscope-motorized scanning stage allows Raman imaging with a resolution of a few hundred nanometers to probe large areas of the sample and to explore the distribution of hot spots. Again, depending on the substrate, large variations in the SERS signal intensity can be expected when scanning over the metal surface, even for a uniform sample distribution, especially in the case of randomly distributed surface roughness. When planning a SERS experiment, particular care and planning is needed concerning the exciting laser wavelength, which is going to produce the best results. This will be dictated by the used SERS substrates and more specifically by the wavelength position of the LSP resonances characterizing the plasmonic substrate. Ideally, the exciting wavelength should match or be close enough to the sample extinction maximum for strong plasmon excitation. As guidelines, green laser light is better suited for silver and copper and for small spherical nanoparticles, while red and infrared sources will provide stronger SERS with gold, high aspect ratio nanoparticles (nanorods) and nanogaps. Aluminum and other unusual SERS materials would require UV excitation.

5.2 Tip-Enhanced Raman Spectroscopy

Tip-enhanced Raman spectroscopy (TERS) resolves many of the limiting aspects of SERS while retaining similarly high enhancement factors and overall sensitivity. The original idea behind TERS was already provided by Wessel in 1985 [29], several years before any real TERS experimental realization in 2000. To extend the SERS effect to surfaces other than the roughened Ag, Au, or Cu substrates or even to study molecules adsorbed on a smooth metal surface, the electromagnetic enhancement could be externally delivered by an illuminated individual Ag or Au nanoparticle, with the associated enhanced field, acting as a single external hot spot, placed above the surface at a very short separation from it, less than 10 nm. This approach has a very attractive feature; if the external hot spot could be precisely moved, scanned over the surface, then the whole sample surface could be probed by the same hot spot, allowing for a 2D mapping of the Raman signal intensity due to the molecular adlayer.

The difficulty of this experiment was indeed to fabricate and place the nano-object within only a few nanometers distance from the surface and to keep it at this constant distance during the scanning. Wessel's idea indeed preceded in time the advent of the scanning probe microscopies. The sharp metal tip of a scanning tunneling microscope (STM) or an atomic force microscope (AFM) perfectly realizes the original intuition, with the only requirement that the tip material is Ag or Au. Actually, TERS (as this technique was much later named) stays at the cross-point between the efforts of two scientific communities: the SERS/plasmonic community and the nearfield community. Indeed, in the 1990s, several attempts were made to extend the SNOM technique, primarily used for studying the fluorescence signal of the sample with sub-diffraction-limited resolution, in conjunction with Raman spectroscopy. These Raman–SNOM studies were hampered by the very low cross section of the Raman signal and by the limited amount of (nearfield) light that could be brought to interact with the sample through the sub-micrometer SNOM aperture. Some authors, while managing to obtain nearfield Raman maps adopting the aperture SNOM configuration, concluded that routine imaging was precluded except under favorable cases [30]. Still in the nearfield community, aperture-less SNOM, where a sharp solid tip replaced the aperture probe, was introduced to improve the nearfield spatial resolution beyond the typical values, which were limited by the physical size of the aperture (~100 nm) [31]. Indeed, silicon tips present apexes as low as 10 nm or even less (the same technology used to fabricate AFM probes). This aperture-less approach strongly depends on the ability of the tip material to scatter the incident light and generate an electromagnetic nearfield, which can interact with the sample. Since silicon has a poor visible light scatter, it was proposed to rather use Au or Ag to increase the nearfield coupling of the incident light with the sharp tip. The tip apex represents a nano-antenna, mediating between the incident far-field and the nearfield. Eventually, the excitation of LSP resonances in Au or Au leads to a very efficient interaction and to the creation of strong local fields near the apex, sufficient to study also weaker optical processes in the sample, as Raman scattering (i.e., TERS).

Since the LSP excitation and electromagnetic enhancement in TERS are mostly associated with the external tip, Raman investigations are no more limited to the molecules adsorbed on coinage metals and virtually any substrate (smooth or rough) can be studied, even molecules dispersed on glass. In principle, and as long as the tip does not enter in contact with the molecular layer, there is a physical decoupling between the two SERS mechanisms: The chemical effect is still associated with the interaction of the molecules with the supporting surface, but the electromagnetic enhancement is now associated with the tip (although, as later discussed, it is not completely true and the substrate participates in defining the magnitude of the electromagnetic effect). Most importantly, TERS brings a whole new dimension to SERS: 2D imaging with nearfield optical resolution. In classical, confocal Raman (far-field) imaging, the best attainable optical resolution is dictated by Abbe's principle (corresponding to ~300 nm if green light and a high numerical aperture objective are employed). In TERS, the exciting source is the nearfield associated with the LSP excitation at the tip. In the very first approximation, the area investigated by TERS has similar or smaller dimensions than the tip apex (radius < 30 nm), which is much smaller than the laser spot size used to externally illuminate the tip. Indeed, only the molecules located directly underneath the (scanning) tip apex profit from the tip enhancement. It can be useful to visualize the illuminated tip as a nano-torch (of nearfield light) brought very close to the sample surface. Additionally, since the vertical position and horizontal movements of the tip are controlled by the piezo elements and electronics of a scanning probe microscope (STM or AFM), the surface topography can be recorded as well (at the same time or before/after the TERS mapping, depending on the implementation). TERS is very much valued since it adds chemical sensitivity to the otherwise "chemically blind" scanning probe microscopies; due to the nearfield optical resolution, the surface topography and chemical composition of the surface are probed with comparable nanometer resolution, which is essentially limited in both cases by the size of the tip apex.

The TERS experimental setup is a combination of a scanning probe microscope (AFM or STM) and a Raman confocal microscope operating in the back-scattering mode. The laser light has to be (re-)directed to illuminate the tip apex. Optimal illumination with tight focusing and efficient collection (short optical paths are preferred) are essential to TERS. There are two main TERS optical configurations (Fig. 5.5): The tip can be illuminated from the bottom, through a transparent substrate (such as a glass slide or very thin metal films), or the tip can be illuminated from the side, which allows opaque samples to be studied as well. The main advantage of the bottom illumination is that very high NA objectives with short focal distance can be used, facilitating tip alignment and providing large collection efficiency while illuminating a very small circular area. Objectives with longer focal distance and minor NA are required for side illumination. In this case, the main advantage is that the incident electric field can be chosen to have a large component along the tip axis and across the nanogap formed between the apex and the surface, which forms the TERS hot spot. STM is more proficient than AFM in maintaining a stable 1–2 nm gap between tip and substrate, and indeed the most impressive TERS results (enhancement factors and lateral resolutions) have been obtained in the STM mode using an Au substrate (to underline the presence of an Au or Ag substrate opposite to the tip; this configuration is sometimes specified as gap-mode TERS). On the other hand, STM is limited to conductive substrates, while AFM can operate on any substrate, even biological objects, employing Ag- or Au-coated Si tips. A further bottleneck of AFM-TERS is the fabrication of reliable and wear-resistant tips. Especially in the contact mode, the metal coating is very easily damaged, while in the non-contact or tapping mode, the tip oscillates most of the time too far from the surface, providing only moderate enhancement. Recently developed AFM hybrid modes (originally introduced for force mapping) have proved to yield improved AFM-TERS performances. It has become somewhat conventional to dedicate bottom illumination and AFMs to investigate biological samples, which can be inspected earlier under white light or using other fluorescence-based techniques profiting of the inverted microscope, while STM and oblique illumination are preferred in material science and molecular adsorption studies on metal surfaces. Finally, AFM-TERS tips and STM-TERS tips are fabricated with very different protocols and different degrees of reproducibility; STM tips are easier and cheaper to fabricate.

The substrate also plays an important role in TERS. Stronger TERS effects have been observed for molecules located in the nano-cavity formed by the Au (Ag) apex and an Au (or Ag) surface. Indeed, when the nano-cavity formed by the tip apex and the substrate, separated by only a few nanometers, is illuminated, the dielectric constants of both materials contribute to the electromagnetic field standing in the gap. Quite understandably, a free electron metal surface will actively contribute to the optical resonance(s), while a dielectric surface (glass or silicon) will rather behave as a passive component, dampening the electron oscillations of the tip. From a classical electromagnetic point of view, the field generated by the surface electron oscillations upon excitation of the plasmon dipolar mode of a sphere (resembling the apex) can be reasonably well associated with the electric field generated by a single dipole located at the center of the sphere and with an effective charge related to the sphere surface charge. When the "tip dipole" is placed a few nanometers above a conductive surface, mirror charges are created. The superposition of the dipole and image dipole fields generates a stronger field [32].

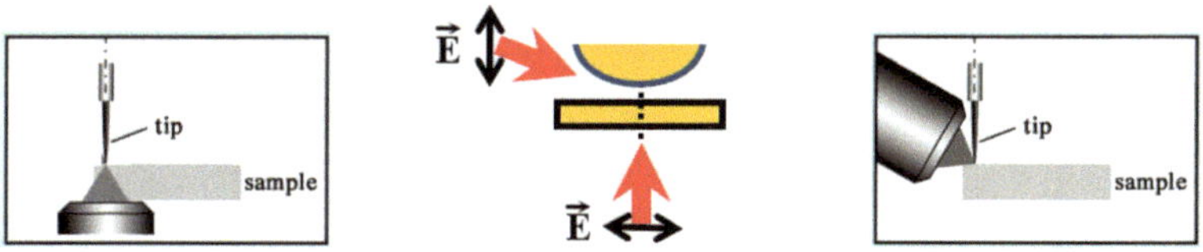

Figure 5.5 The two main optical configurations adopted in TERS: On the left, the tip is illuminated from below by a microscope objective located below the sample (bottom illumination); on the right, the objective is located above the sample and the tip is illuminated from the side (oblique illumination). The scheme in the center displays the incident wave vector (red arrows) with respect to the sample and tip apex (in yellow) for the two configurations and the orientation of the associated electric fields (black double-headed arrows).

A number of review articles have presented the major advancements in the development of TERS and several representative applications [33–36]. In the following, only a few examples are shown to discuss the main TERS features.

An example of TERS capability in enhancing the Raman signal of molecular layers is shown in Fig. 5.6 [37]. In this experiment, MGITC dye molecules are adsorbed at an atomically smooth Au(111), which is mostly SERS inactive due to lacking any surface roughness (no electromagnetic enhancement). The black trace is the resonance Raman spectrum recorded in the absence of the TERS tip (the tip retracted several microns above the surface). Some weak Raman bands can be already recognized due to the resonance Raman effect of the dye. Adsorption on the metal surface efficiently quenches the concomitant fluorescence, a quite useful feature of TERS on metal substrates. The red trace is recorded when illuminating the same spot (i.e., the same number of molecules, actually exactly the same molecules), but this time the TERS Au tip is in tunneling contact with the surface at a distance of ~1 nm (defined by the set tunneling voltage and tunneling current). As the tip is lowered to contact, it enters the laser spot and is thus illuminated. The TERS effect operates on the molecules located right below the apex.

To evaluate the Raman signal enhancement factor, the signal contrast is first calculated. The contrast C is simply the ratio of the integral intensity of a particular Raman band, normally the most intense, recorded with and without the tip. Then, the following formulas are used:

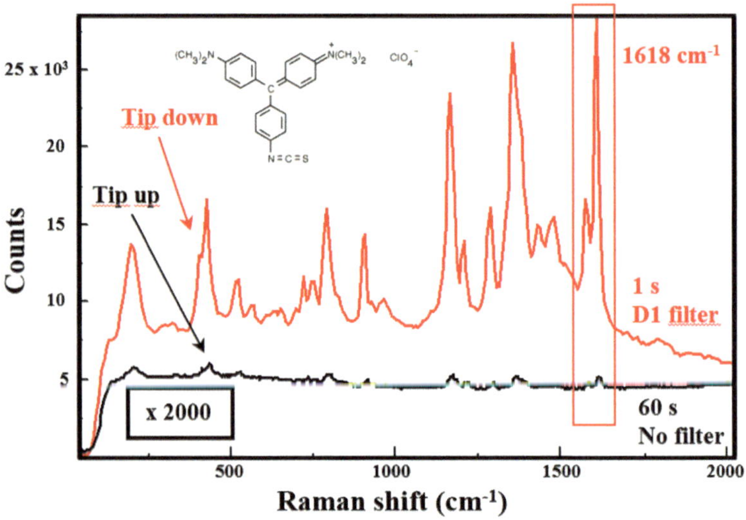

Figure 5.6 Comparison between the Raman spectrum of the resonant dye malachite green isothyocianate (MGITC) adsorbed on atomically smooth Au(111) when the TERS Au tip was retracted far above the sample and not illuminated (tip up, black trace) and the TERS spectrum was recorded while illuminating the same spot of the surface but with the TERS tip in tunneling contact and illuminated (tip down, red trace). The TERS spectrum was recorded in 1 s, and with a 10 times lower laser intensity, the tip-up spectrum was further multiplied by a factor 2000. The strongest band at 1618 cm^{-1} was used to evaluate the contrast between the Raman intensities tip down/tip up, giving a value > 10^4. Adapted from Ref. [37].

$$C = \frac{I_{\text{tip down}}}{I_{\text{tip up}}} = \frac{I_{\text{nearfield}} + I_{\text{far-field}}}{I_{\text{far-field}}} \tag{5.2}$$

$$EF = \frac{\left(I_{\text{nearfield}} \Big/ A_{\text{nearfield}} \right)}{\left(I_{\text{far-field}} \Big/ A_{\text{far-field}} \right)} = (C-1)\frac{A_{\text{far-field}}}{A_{\text{nearfield}}} = (C-1)\frac{R_{\text{spot}}^2}{R_{\text{tip}}^2} \tag{5.3}$$

A is the focal area defined by the NA of the objective and the employed wavelength, and R is its radius. To obtain the enhancement factor, the tip-down and tip-up intensities should be normalized by the number of molecules participating in the respective spectra; for a uniformly distributed molecular layer of Raman scatterer, the far-field and nearfield areas can be considered. The far-field is the laser spot area (having usually circular spots of radius R_{spot} ~0.5 μm and bottom illumination); all the illuminated molecules participate in the "unenhanced" far-field signal. The nearfield area is the small disk region below the tip, having a radius approximated by the physical radius of the apex (R_{tip} ~30 nm). Only a very small subset of all the illuminated molecules contributes to the TERS signal, but this smaller number of molecules provides a much stronger Raman signal than all the illuminated molecules thanks to the tip enhancement. In the gap-mode TERS, where the apex and metal substrate form a nano-cavity, the effective lateral extension (localization) of the squeezed plasmon mode is $L = (2R_{\text{tip}}D)^{1/2}$ [38], where D is the distance from the flat surface to the bottom of a sphere having a radius R_{tip} and the TERS radius can be approximated to $L/2$. Note that the tip-down spectra are an incoherent sum of the far-field and nearfield signals; in a number of cases, the far-field is extremely weak compared to the TERS intensity or altogether absent and the tip-up spectra consisting only of a baseline due to the substrate; in this case, formula 5.3 can be simplified, removing the factor one. Similar formulas can be used for SERS if the number and size of the contributing hot spots are known, as it is indeed well known for TERS or for certain lithographically fabricated SERS substrates. Thus, the estimations on the absolute Raman signal enhancement factor in SERS are more varied and based on assumptions, than in the case of TERS. Generally, apart from the very different intensity level, a TERS spectrum mostly resembles the unenhanced, far-field spectrum of the adsorbed molecules. In some instances, band narrowing and maximum shifts are to be ascribed to TERS, probing a much smaller number of molecules with less averaging of different surface sites and molecular configurations.

While Fig. 5.6 shows the signal enhancement, the first advantageous TERS feature, Fig. 5.7 nicely shows the improvement in lateral resolution when passing from far-field Raman imaging to nearfield Raman imaging [39]. The same isolated single-wall carbon nanotube (SWCNT) is probed: The left image corresponds to a classical confocal Raman image (no illuminated tip), specifically to the intensity of the 2D band. SWCNTs provide strong Raman scattering due to resonant effects; thus, even a single isolated tube yields sufficient signal without any other external enhancement. Anyhow, the limit of far-field imaging is evident considering that SWCNTs are almost 1D cylindrical objects with typical lateral size of ~1 nm. In the confocal Raman image, the tube is seen with a much larger width of ~350 nm. This apparent width is due to the convolution of the real width (almost negligible) of SWCNTs and the diffraction-limited spot size. In the TERS image (same optically scanned region, but with the tip brought in contact with the sample and illuminated), the apparent width is only ~14 nm, since now the real width is convoluted with the size of the nearfield (TERS) radius, which also defines the optical resolution in this particular experiment.

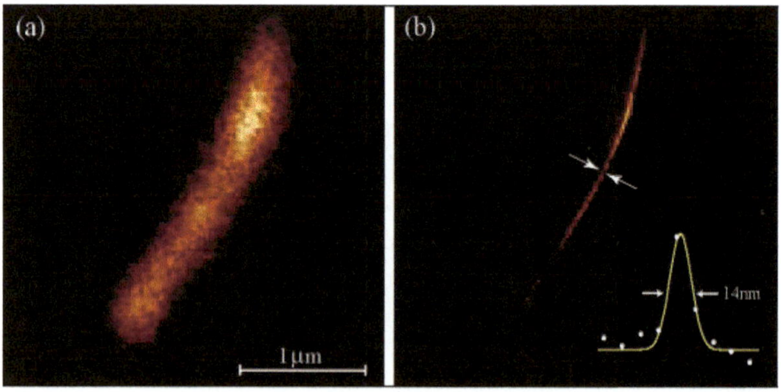

Figure 5.7 Raman confocal image (left) and TERS image of the same SWCNT deposited on a glass slide. The color reflects the intensity of the SWCNT 2D band at 2640 cm^{-1}: A higher intensity is represented by a brighter color. Reproduced with permission from Ref. [39].

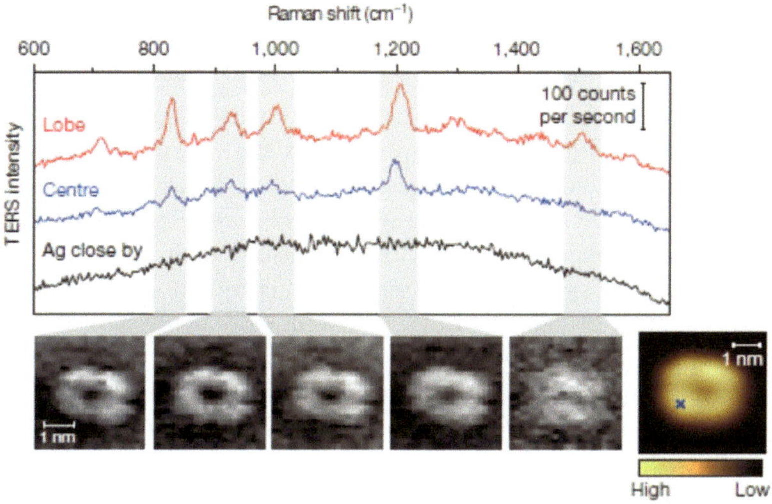

Figure 5.8 TERS spectra recorded from a single H2TBPP molecule on Ag (111). The blue and red traces refer to two different positions of the tip on the molecule; the black trace is relative to the tip positioned on the bare Ag surface about 1 nm laterally away from the molecule. The bottom row shows 2D TERS images obtained integrating the intensity of the greyed-out bands. Some pixilation is still visible corresponding to the finite step size (~0.16 nm) used to obtain these 2D intensity maps (23 × 23 pixels). On the bottom right an UHV-STM topography image of an individual H2TBPP molecule adsorbed flat at low temperature is shown. Reproduced with permission from Ref. [40].

Some reports demonstrated that the TERS resolution can even reach below the nanometer level [40]. The spectra and Raman images in Fig. 5.8 refer to a single porphyrin molecule deposited on an Ag substrate. A low-temperature STM under UHV conditions was used to minimize molecular diffusion on the flat surface and achieve atomic resolution in the STM topography images to be compared with the TERS images. The gap-mode STM configuration (Au tip/ Ag surface) is compulsory to have the highest field enhancement and to be sensitive to a single molecule. Still, it was further required to finely tune the LSP resonance of the tip-sample cavity with the energy of the Raman-scattered photons by controlling with Å precision the tip to sample distance. Note that the quadrilobe structure of individual porphyrin molecules is distinguished also in the topographic STM images; on the other hand, the TERS images on the bottom are indeed *optical images* recorded on a very narrow wavelength window (corresponding to photons Raman scattered by a particular molecular vibration) within the visible range. The resolution is down to only a few Å in both kinds of images.

Still, this submolecular optical resolution is not fully compatible with the common TERS image in which the lateral resolution is associated with the physical size of the apex and with the external electromagnetic field enhancement associated with the LSP excitation in nanostructure with dimensions 5 to 30 nm. A full rationalization of TERS at an atomic level is still missing, and more experimental evidence must be gathered. More advanced concepts have been invoked [41], some of which common to SERS: For example, smaller structures (with nanoscale roughness on the larger spherical apex) can be responsible for even higher field strength and field localization and/or for an additional "internal" enhancement (reminiscent of the chemical effect) present only for specific interactions between the tip and molecule at very short distances, possibly involving molecular tunneling and specific hybrid states of the molecule and clusters of metal atoms.

5.3 SERS/TERS Applications in Plasmon-Mediated Chemical Reactions

SERS and TERS are analytical techniques useful to chemically identify molecules surrounding the illuminated metallic nanoparticles supporting the LSP optical excitation. By adopting well-planned procedures, these techniques can be used to monitor in time the chemical reactions taking place at the metal surface, ideally as the spectral signature of the reactants transforms into the spectral signature of the products, or more generally by detecting spectral changes indicative of a surface process taking place. In these instances, the illuminated nanoparticles play the double role of Raman signal enhancer (sensing the environment) and promoter of chemical changes (modifying the environment). In both cases, the efficient excitation of LSP modes is the key step. To disentangle the two roles, two general strategies exist: In a number of reports, an illumination intensity threshold exists for triggering surface chemical process, likely related to the number of hot electrons generated and transferred to the adsorbate or to the required temperature increase associated with the thermoplasmonic effect. The SERS characteristics can be first probed when illuminating below the threshold, to avoid or minimize the modification of the original molecular structure; then, laser pulses with intensity above the threshold value are applied, and finally the metal surface is again probed with reduced laser power. A humble precursor of this approach is the "burning" of a molecular dye layer under strong illumination. The molecules adsorbed at a metal substrate provide a recognizable pattern of Raman bands when illuminated at medium low intensity. If high laser powers are used, very strong and mostly random SERS/TERS fluctuations appear, signaling the tumultuous formation of many different and short-lived chemical species or intermediates (which can be difficult to identify due to band overlaps), until only amorphous carbon traces are left on the surface. This complex and very fast surface chemistry (involving also oxygen from air) is the result of the LSP excitation and the generation of highly energetic electrons near the surface. An example predating the interest in plasmon-enhanced chemical reaction is [42], which mainly focuses on the evaluation of the hottest SERS sites, their distribution and contribution to the overall SERS signal using an Ag film on nanospheres (FON) (Fig. 5.9).

Molecules at the hottest sites are "damaged" when the local field $gE_{\text{incident}} \geq E_{\text{th}}$, where E_{th} is the threshold field, depending on the type of molecule, the wavelength, and the pulse duration and g is the field enhancement factor due to the LSP excitation. The molecules located where g is largest "burn" first. Spectrally, there is a loss of intensity (3050 cm^{-1} phenyl ν_{CH}) or broadening (near 1550 cm^{-1}) of some of the original bands and the appearance of new features (near 2100 cm^{-1}, carbonyl stretch resulting from the interaction of the photoproduct with oxygen). Anyhow, the used pulses (with energy in the range of 3 to 20.10^7 V/m) were not intense enough to damage the substrate and/or modify the magnitude of the SERS effect.

More recently, studying the SERS/TERS spectral variations for increasing laser fluency, Sczerbinsky et al. [43] focused on the molecular cross-linking (formation of intermolecular C–C bonds) in a series of model aliphatic and aromatic thiol self-assembled monolayers (SAMs). The products are equaled to those observed upon X-ray or electron beam irradiation, supporting reaction mechanisms involving hot electrons with energies up to a few eVs above the metal Fermi level generated upon visible light illumination. These are assimilated to the secondary electrons scattered by the metal substrate upon nominally much more energetic X-ray irradiation. Ruling out of thermal mechanisms is based on comparatively analyzing the products under thermal programmed desorption mass spectrometry (TPD-MS). These results support an alternative explanation for the origin of the well-documented (and often limiting) SERS/TERS spectral fluctuations, toning down the role of local temperature increase upon the illumination of the SERS-active substrate ("burning" of the molecular adlayer in the presence of ambient oxygen) and designate the surface release and further recombination of reactive molecular fragments, such as radicals formed in the presence of hot electrons, in an induced desorption process.

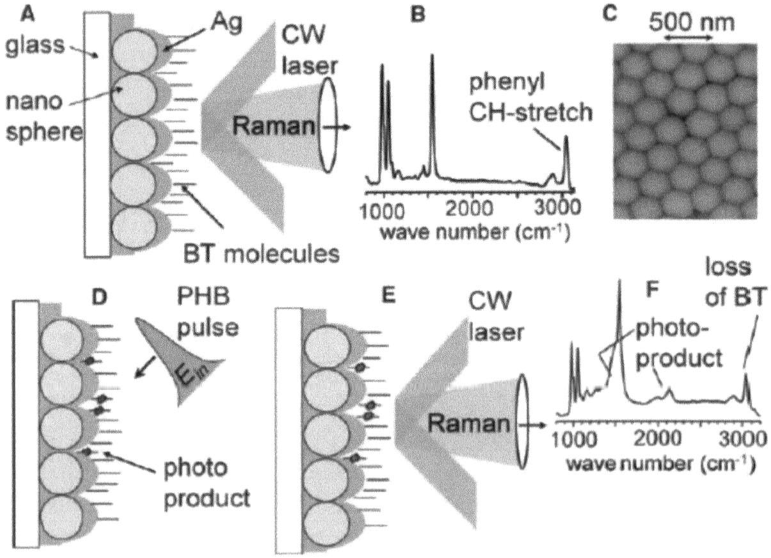

Figure 5.9 In this study, the original molecular layer (benzenethiol molecules BT, represented by straight lines) adsorbed on the Ag substrate is analyzed by SERS (upper row). Intense laser pulses of increasing power are then used to illuminate the sample, and photoproducts are formed and subsequently analyzed by SERS (bottom row). The spectra show the presence of new species, mostly localized in the gap between adjacent spheres (small black circles), where electromagnetic fields are stronger. Reproduced with permission from Ref. [42].

A second approach to study chemical reactions by means of SERS/TERS is based on adopting a dual-wavelength illumination: One wavelength is used to promote the chemistry, and a different wavelength is used to excite the SERS/TERS effect. In principle, this approach allows to follow in real time the chemical transformations. The two chosen wavelengths should fall within the same broad LSP resonance (up to ~100 nm wide) for the NP to actively participate in both mentioned roles. In this respect, note that the sites of maximal SERS effect might not be necessarily the same sites responsible for the strongest chemical reactivity. This is especially true for the thermoplasmonic effect, which mostly relies on the effects of the vigorous electron oscillation within the metal body (ultimately, joule heating) as opposed to the generation of strong electromagnetic fields outside the nanoparticles necessary for SERS [44]. Concerning reaction mechanisms involving hot electrons, not only their generation should be considered, which can be more or less related to the local field intensity, but also their lifetime and transfer to the adsorbate, which can also be site dependent. The field of plasmon-assisted chemical reaction is still dawning, and a number of experimental and theoretical issues need more investigations, accepting that concepts valid for a given chemical system ("reaction") may not be transferable to a different system. The difficulty to generalize also lies on the specific way a molecule interacts with the LSP exciton.

A molecular system that has been often investigated by both SERS and TERS is the dimerization of 4-aminothiophenol (4ATP, terminal amino group) and 4-nitrothiophenol (4NTP, terminal nitro group) SAMs to the same dimercaptoazobenzene (DMAB) product, presenting an N=N double bound, by oxidation and reduction, respectively [45]. All those molecules interact strongly with the metal surface via the S atoms, display large Raman cross sections (mostly due to the aromatic skeleton), and present rather clean spectral signatures. For the oxidation process involving two adjacent 4ATP molecules, the appearance of different, additional bands was initially not recognized due to newly formed chemical species, but rather to an exceptional surface enhancement of otherwise weak bands of b2 symmetry. Tian's group and Sun's group [46, 47] advanced the molecular dimerization process, depicted in Fig. 5.10, and the bands at 1140 cm^{-1}, 1388 cm^{-1}, and 1438 cm^{-1} were assigned to δ_{CH} and ν_{NN}

modes from the dimer at the Ag roughened surface. The scissoring of the double bond of the surface-attached DMAB molecules (thus a reduction back to two molecules of 4ATP) is also possible, and in this reverse case, a participation of hot electrons was invoked with H_2 or H_2O acting as the hydrogen source [48, 49]. Alternating proper oxidizing and reducing atmospheres, the molecular dimerization and scission may be cyclically repeated.

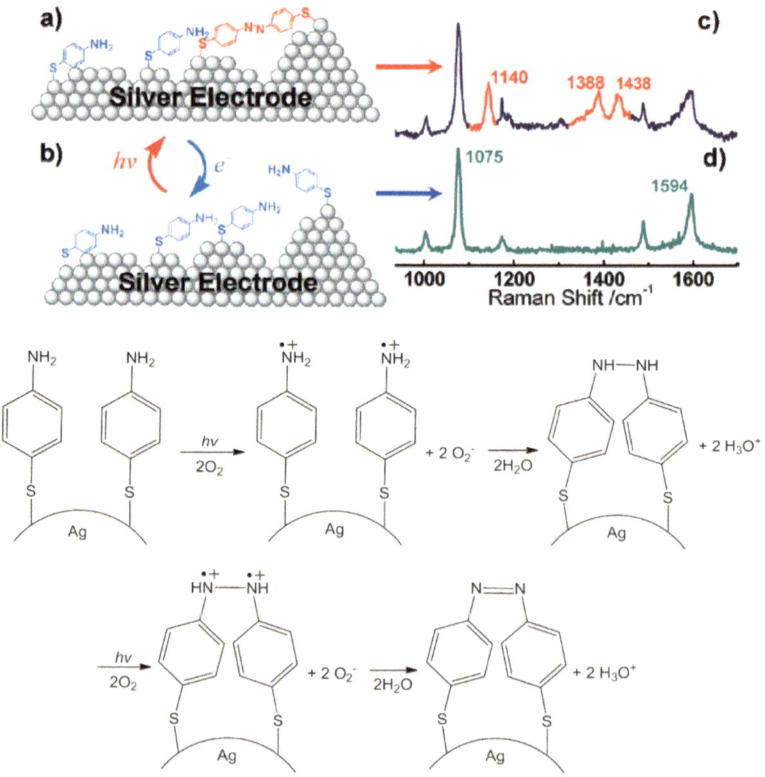

Figure 5.10 On the left, the dimerization of 4ATP (bottom row) to DMAB (upper row) and reverse process. The additional DMAB bands are evidenced in red. Reproduced with permission from Ref. [46]. On the right, the mechanism proposed by Xu et al. for the conversion of 4ATP to DMAB under irradiation and plasmonic heating, further involving oxygen as the electron acceptor (oxidant) and water molecules for the deprotonation of 4ATP (basic condition accelerates the process). Reproduced with permission from Ref. [48]. Note that the NN bond breaking follows a different mechanism involving hot electrons.

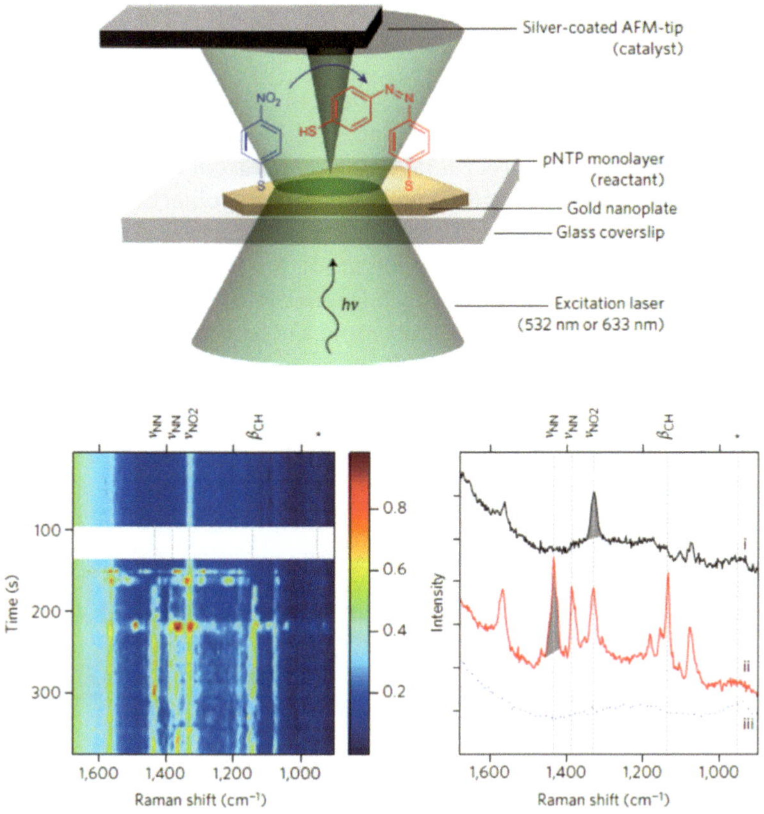

Figure 5.11 (Left) Schematics of the TERS setup used to monitor catalytic processes on metal surfaces at the nanoscale, in particular the reduction of 4NTP to DMAB. (Right) Time evolution of the TERS intensity at 633 nm excitation (waterfall plot) at 633 nm excitation before (top region, no change in the spectral feature of 4NTP) and after illumination at 532 nm (below white band). Two individual spectra from the waterfall plot: Spectrum (i) is taken at 90 s (original 4NTP SAM) and spectrum (ii) at 265 s. Reproduced with permission from Ref. [52].

The reduction process of two adjacent 4NTP molecules to DMAB on Au, Ag, and Cu SERS substrates illuminated at different wavelengths is documented in Refs. [50, 51]. The spectral evidence is the disappearance of the ν_s signal of the nitro group at 1331 cm^{-1} and the appearances of new bands at 1387 cm^{-1} and 1432 cm^{-1} having strong contributions from the nitrogen double bond in DMAB. Four hot electrons stemming from the LSP decay are required for the reduction of two nitro groups and the formation of a DMAB molecule. This hot-electron-driven process has been also investigated by TERS [52, 53]. In Ref. [52], the authors use as semitransparent substrate (bottom illumination AFM-TERS) chemically synthesized nano-plates of Au dispersed on glass for the tethering of the 4NTP SAM and to profit from the gap mode also under bottom illumination with a large NA objective (Fig. 5.11). The nano-plates present an ultrasmooth surface comparable to the 111 terraces of bulk crystalline Au. Furthermore, in this work, the double wavelength approach is used, with 532 nm light used to trigger the photocatalytic activity and the less invasive (for this gap configuration) 633 nm light only for monitoring purposes.

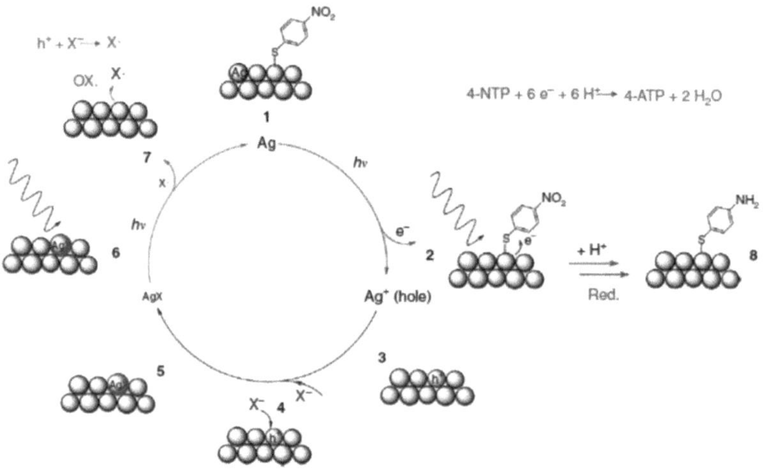

Figure 5.12 Scheme of the LSP-assisted reduction of 4NTP to 4ATP on laser-illuminated AgNP in the presence of AgCl. Holes h^+ are left on the NP (3), which are captured by halide ions (4), leading to the precipitation of the Ag salt (5), which readily photodissociates under illumination (6). Reproduced with permission from Ref. [54].

An important and general observation, inferred from the reduction of 4NTP to 4ATP, was made by Xie and Schuckler [54]. When using Ag nanoparticle in the presence of AgCl in solution, the reaction proceeds also in the absence of the reducing agent $BH4^-$ ions. The halide ions play a decisive role compensating the holes left on the AgNPs after the hot electron transfer and reduction of 4NTP, via an oxidation half reaction and recycling of the plasmonically active Ag surface. As exemplified in Fig. 5.12, overall six electrons are required to complete the reduction of one 4NTP molecule (one full cycle); the oxidation counterpart (steps from 4 to 7) can be globally written as h^+ (hole left in the AgNP) + Cl-s $\rightarrow$ Cl·. The Cl· radical is actually formed (step 6) from the photodissociation of the poorly soluble AgCl, which precipitates on the NP. Generally speaking, whenever a hot electron mechanism is advanced, the fate of the simultaneously generated hot holes is also important and it can actually limit the whole chemical process, if these just recombine after a short time with the hot electrons: A suitable oxidation half reaction must coexist to sustain the reduction and balance the charge left of the NP. The work of Jain also underlines this point [55].

Tijunelyte et al. [56] employed Au nanocylinders of different diameters fabricated by electron beam lithography (EBL) as a tunable substrate for a thiol-ene "click" chemical reaction. 1,4-butanedithiol, bearing a double-bond terminal group, was initially self-assembled on the Au cylinders. The modified surface was then exposed to thiophenol in the liquid phase in the presence of a radical catalyzer (AAPH, active at temperature above 60°C or under UV illumination) and irradiated by 660 nm laser light (Fig. 5.13). The reaction kinetics were then monitored by SERS in real time with acquisitions every minute following the increase in the intensity of the aromatic ring signal ($\nu_{C=C}$ at 1572 cm^{-1}), as the thiophenol molecules from the solution reacted with the allyl molecules grafted on the Au surface and were thus increasingly immobilized there. With the support of several negative control experiments (for example to exclude spontaneous adsorption of thiophenol on gold by thiol exchange) and by testing with different sizes of cylinders (110 nm, 140 nm, 200 nm in diameter), characterized by LSP resonances at different wavelengths, the authors prove that the illumination of the 110 nm AuNP (having the extinction maximum the closest to 660 nm) accelerates up to 20 times the reaction kinetics and that this effect is related to the LSP excitation. Indeed, the reaction time upon illumination at the used fixed wavelength is strongly dependent on the LSP position. Concerning the operating mechanism(s), both a thermoplasmonic effect, sufficient to raise the local temperature above 60°C, and a broadening or shifting of the electronic levels of the AAPH catalyzer (normally requiring UV excitation to initiate the radical cycle) upon interaction with the gold surface, allowing the absorption of less energetic photons in the visible region, are advanced.

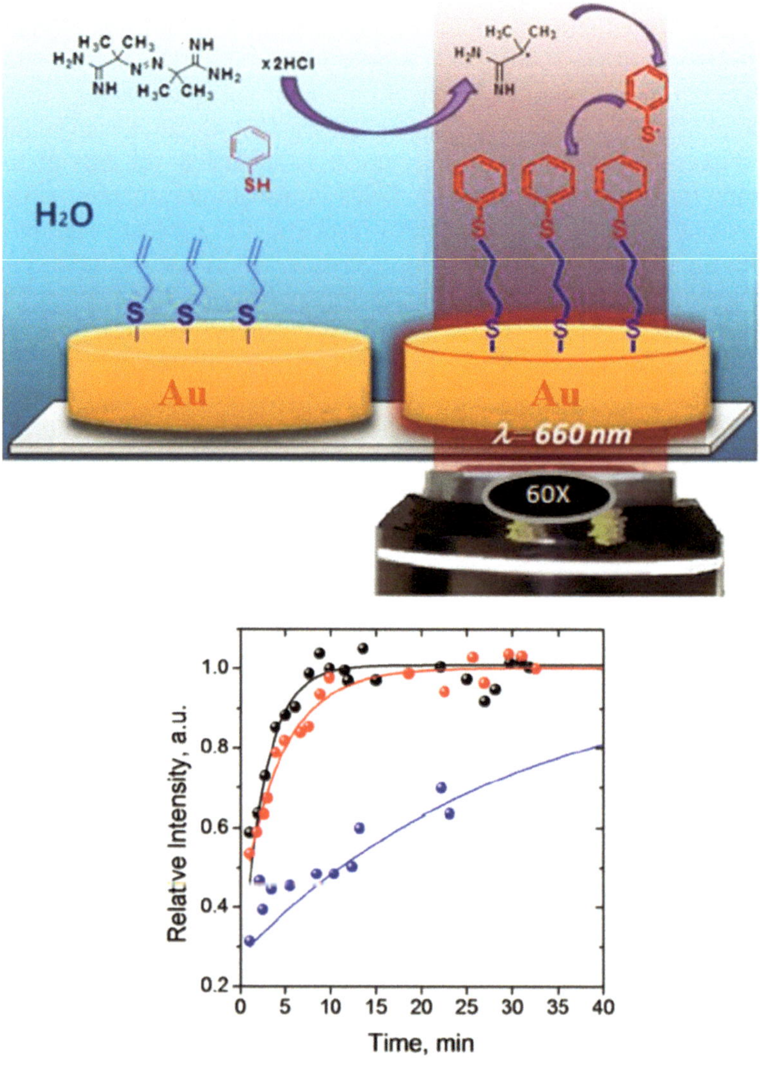

Figure 5.13 Schematics of the click-chemistry reaction taking place at the surface of laser-illuminated Au nanocylinders. The radical initiator (2,2′-azobis(2-methylpropionamidine) dihydrochloride (AAPH)) forms a thiyl radical, which then reacts with the terminal C=C bond of the surface molecules. On the right, normalized SERS intensities (1572 cm^{-1}) versus reaction time for the 110 nm (black dot), 140 nm (red dot), and 200 nm (blue dot) nanocylinders. Reproduced with permission from Ref. [56].

In the following last example, Raman spectroscopy is not used, but the entire setup and the used probes are exactly those employed for gap-mode TERS using an STM (under UHV environment) [57]. The authors studied in real time by high-resolution STM, the photodissociation of the dimer molecule $(CH_3S)_2$ taking place at single-molecule level within the plasmonic cavity formed between a tunneling Ag tip and an Ag surface. The main message of this work is to demonstrate that this molecular dissociation proceeds by *direct* intramolecular excitation induced by the LSP (energy directly transferred from the LSP to the molecule, more precisely to a LUMO hybridized with the metal states but still presenting a large $\sigma^*_{S\text{-}S}$ character) and not by an *indirect* LSP mechanism involving LSP decay, generation of hot electrons, their transfer to form a transient negative ion (TNI), which finally decays to highly excited vibrational states (leading to dissociation) (see Fig. 5.14) as advanced for other plasmon-assisted photo-dissociations (O_2 and H_2). The evidence is based on two series of experiments: First, plotting the wavelength dependence of the plasmon-induced chemical reaction reveals more similarities to the "normal" photodissociation process (no tip, no LSP excitation), although taking place in a shifted energy range due to the involvement of LSP excitation. No energy threshold effect is observed, which would be expected for the transfer of hot electrons. They further compared real-time STM results for the plasmon-induced reaction (hypothesis of direct mechanism) and the same process taking place in the dark but under inelastic electron tunneling, in which the energy of the electrons can be controlled by the applied tunneling voltage and are shuttled through the nano-cavity, essentially "creating" ballistic hot electrons without illumination and excitation of the LSP modes (indirect pathway). The degree of hybridization between the molecular orbitals and the metal states (or more generally the electronic interaction of the molecule with the metal defined by their MO structure and DOS) plays a crucial role in the mechanism underlying the plasmon-induced chemical reactions.

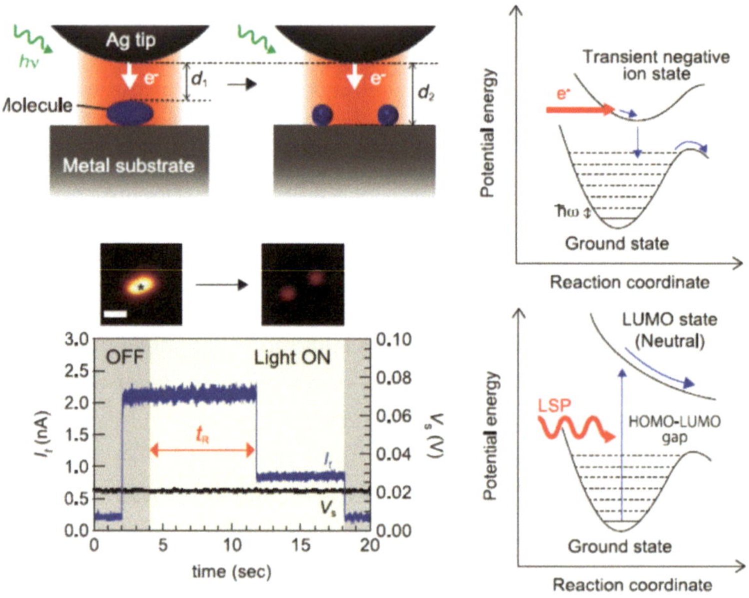

Figure 5.14 (Above left) Schematic of the molecular photodissociation process induced by illuminating the nano-cavity formed between the tunneling Ag tip and the metal substrate. The red area symbolizes the localization region of the excited LSP gap mode. In the middle, high-resolution, UHV-STM images of an individual $(CH_3S)_2$ dimer, before and after dissociation (the tip was located at the position marked with a cross, the scale bar is 0.5 nm). (Bottom left) The tunneling current under constant height STM operation mode recorded during the photodissociation of a single molecule; at $t = 4$ s, laser light is switched on and at $t = 18$ s switched off. The current drop at $t = 12$ s is due to the change in the tunneling resistivity (i.e., tunneling distance) when the molecule is dissociated. The delay time t_R depends on the local field intensity: stronger LSP excitation (for smaller gap distances), shorter t_R. On the left, the two discussed mechanisms involving the hot electron transfer to form a TNI (indirect mechanism) and the direct pathway. Reproduced with permission from Ref. [57].

References

1. Fleischmann, M., P. J. Hendra, and A. J. McQuillan, Raman spectra of pyridine adsorbed at a silver electrode. *Chemical Physics Letters*, 1974. **26**(2): 163–166.

2. Jeanmaire, D. L. and R. P. Van Duyne, Surface Raman spectroelectrochemistry: Part I. Heterocyclic, aromatic, and aliphatic

amines adsorbed on the anodized silver electrode. *Journal of Electroanalytical Chemistry and Interfacial Electrochemistry*, 1977. **84**(1): 1–20.

3. Albrecht, M. G. and J. A. Creighton, Anomalously intense Raman spectra of pyridine at a silver electrode. *Journal of the American Chemical Society*, 1977. **99**(15): 5215–5217.
4. Pemberton, J. E., et al., Surface enhanced Raman scattering in electrochemical systems: The complex roles of surface roughness. *Applied Surface Science*, 1988. **32**(1): 33–56.
5. Wetzel, H., H. Gerischer, and B. Pettinger, Surface enhanced Raman scattering from silver-halide and silver-pyridine vibrations and the role of silver ad-atoms. *Chemical Physics Letters*, 1981. **78**(2): 392–397.
6. Kudelski, A. and J. Bukowska, Charge-transfer contribution to surface-enhanced Raman scattering and surface-enhanced resonance Raman scattering of dyes at silver and gold electrodes. *Chemical Physics Letters*, 1996. **253**(3): 246–250.
7. Wu, D.-Y., et al., Electrochemical surface-enhanced Raman spectroscopy of nanostructures. *Chemical Society Reviews*, 2008. **37**(5): 1025–1041.
8. Yang, K.-H., Y.-C. Liu, and C.-C. Yu, Enhancements in intensity and stability of surface-enhanced Raman scattering on optimally electrochemically roughened silver substrates. *Journal of Materials Chemistry*, 2008. **18**(40): 4849–4855.
9. Gao, P., et al., Surface-enhanced Raman scattering at gold electrodes: Dependence on electrochemical pretreatment conditions and comparisons with silver. *Journal of Electroanalytical Chemistry and Interfacial Electrochemistry*, 1987. **233**(1): 211–222.
10. Kudelski, A., et al., Characterization of the copper surface optimized for use as a substrate for surface-enhanced Raman scattering. *Vibrational Spectroscopy*, 1998. **16**(1): 21–29.
11. Jacobse, L., et al., Correlation of surface site formation to nanoisland growth in the electrochemical roughening of Pt(111). *Nature Materials*, 2018. **17**(3): 277–282.
12. Leung, L. W. H. and M. J. Weaver, Extending surface-enhanced Raman spectroscopy to transition-metal surfaces: Carbon monoxide adsorption and electrooxidation on platinum- and palladium-coated gold electrodes. *Journal of the American Chemical Society*, 1987. **109**(17): 5113–5119.

13. Tian, Z.-Q. and B. Ren, Adsorption and reaction at electrochemical interfaces as probed by surface-enhanced Raman spectroscopy. *Annual Review of Physical Chemistry*, 2004. **55**(1): 197–229.
14. Mrozek, M. F., Y. Xie, and M. J. Weaver, Surface-enhanced Raman scattering on uniform platinum-group overlayers: Preparation by redox replacement of underpotential-deposited metals on gold. *Analytical Chemistry*, 2001. **73**(24): 5953–5960.
15. Maver, U., et al., Recent progressive use of atomic force microscopy in biomedical applications. *TrAC Trends in Analytical Chemistry*, 2016. **80**: 96–111.
16. Seo, Y. and W. Jhe, Atomic force microscopy and spectroscopy. *Reports on Progress in Physics*, 2007. **71**(1): 016101.
17. Alessandrini, A. and P. Facci, AFM: A versatile tool in biophysics. *Measurement Science and Technology*, 2005. **16**(6): R65–R92.
18. Flater, E. E., et al., Towards easy and reliable AFM tip shape determination using blind tip reconstruction. *Ultramicroscopy*, 2014. **146**: 130–143.
19. Pettinger, B., et al., Surface-enhanced and STM-tip-enhanced Raman spectroscopy at metal surfaces. *Single Molecules*, 2002. **3**(5–6): 285–294.
20. Chang, R. K. and T. E. Furtak, *Surface Enhanced Raman Scattering*, 1982. New York: Plenum Press. viii, 423 p.
21. Le Ru, E. C. and P. G. Etchegoin, *Principles of Surface-Enhanced Raman Spectroscopy: And Related Plasmonic Effects*, 1st ed, 2009. Amsterdam, Boston: Elsevier. xxiii, 663 p.
22. Langer, J., et al., Present and future of surface-enhanced Raman scattering. *ACS Nano*, 2020. **14**(1): 28–117.
23. Pettinger, B., Light scattering by adsorbates at Ag particles: Quantum-mechanical approach for energy transfer induced interfacial optical processes involving surface plasmons, multipoles, and electron-hole pairs. *The Journal of Chemical Physics*, 1986. **85**(12): 7442–7451.
24. Furtak, T. E. and S. H. Macomber, Voltage-induced shifting of charge-transfer excitations and their role in surface-enhanced Raman scattering. *Chemical Physics Letters*, 1983. **95**(4): 328–332.
25. Lombardi, J. R., et al., The effect of molecular structure on voltage induced shifts of charge transfer excitation in surface enhanced Raman scattering. *Chemical Physics Letters*, 1984. **104**(2): 240–247.
26. Otto, A., The 'chemical' (electronic) contribution to surface-enhanced Raman scattering. *Journal of Raman Spectroscopy*, 2005. **36**(6–7): 497–509.

27. Langer, J., S. M. Novikov, and L. M. Liz-Marzán, Sensing using plasmonic nanostructures and nanoparticles. *Nanotechnology*, 2015. **26**(32): 322001.

28. Haynes, C. L., et al., Surface-enhanced Raman sensors: Early history and the development of sensors for quantitative biowarfare agent and glucose detection. *Journal of Raman Spectroscopy*, 2005. **36**(6–7): 471–484.

29. Wessel, J., Surface-enhanced optical microscopy. *Journal of the Optical Society of America B*, 1985. **2**(9): 1538–1541.

30. Webster, S., D. N. Batchelder, and D. A. Smith, Submicron resolution measurement of stress in silicon by near-field Raman spectroscopy. *Applied Physics Letters*, 1998. **72**(12): 1478–1480.

31. Inouye, Y. and S. Kawata, Near-field scanning optical microscope with a metallic probe tip. *Optics Letters*, 1994. **19**(3): 159–161.

32. Pettinger, B., et al., Tip-enhanced Raman scattering: Influence of the tip-surface geometry on optical resonance and enhancement. *Surface Science*, 2009. **603**(10): 1335–1341.

33. Pettinger, B., et al., Tip-enhanced Raman spectroscopy: Near-fields acting on a few molecules. *Annual Review of Physical Chemistry*, 2012. **63**(1): 379–399.

34. Schmid, T., et al., Nanoscale chemical imaging using tip-enhanced Raman spectroscopy: A critical review. *Angewandte Chemie International Edition*, 2013. **52**(23): 5940–5954.

35. Mauser, N. and A. Hartschuh, Tip-enhanced near-field optical microscopy. *Chemical Society Reviews*, 2014. **43**(4): 1248–1262.

36. Zhang, Z., et al., Tip-enhanced Raman spectroscopy. *Analytical Chemistry*, 2016. **88**(19): 9328–9346.

37. Pettinger, B., et al., Nanoscale probing of adsorbed species by tip-enhanced Raman spectroscopy. *Physical Review Letters*, 2004. **92**(9): 096101.

38. Rendell, R. W. and D. J. Scalapino, Surface plasmons confined by microstructures on tunnel junctions. *Physical Review B*, 1981. **24**(6): 3276–3294.

39. Anderson, N., et al., Nanoscale vibrational analysis of single-walled carbon nanotubes. *Journal of the American Chemical Society*, 2005. **127**(8): 2533–2537.

40. Zhang, R., et al., Chemical mapping of a single molecule by plasmon-enhanced Raman scattering. *Nature*, 2013. **498**(7452): 82–86.

41. Richard-Lacroix, M., et al., Mastering high resolution tip-enhanced Raman spectroscopy: Towards a shift of perception. *Chemical Society Reviews*, 2017. **46**(13): 3922–3944.
42. Fang, Y., N.-H. Seong, and D. D. Dlott, Measurement of the distribution of site enhancements in surface-enhanced Raman scattering. *Science*, 2008. **321**(5887): 388–392.
43. Szczerbiński, J., et al., Plasmon-driven photocatalysis leads to products known from E-beam and X-ray-induced surface chemistry. *Nano Letters*, 2018. **18**(11): 6740–6749.
44. Lalisse, A., et al., Quantifying the efficiency of plasmonic materials for near-field enhancement and photothermal conversion. *Journal of Physical Chemistry C*, 2015. **119**(45): 25518–25528.
45. Ren, X., et al., Recent advances in surface plasmon-driven catalytic reactions. *RSC Advances*, 2017. **7**(50): 31189–31203.
46. Huang, Y.-F., et al., When the signal is not from the original molecule to be detected: Chemical transformation of para-aminothiophenol on Ag during the SERS measurement. *Journal of the American Chemical Society*, 2010. **132**(27): 9244–9246.
47. Fang, Y., et al., Ascertaining p,p′-dimercaptoazobenzene produced from p-aminothiophenol by selective catalytic coupling reaction on silver nanoparticles. *Langmuir*, 2010. **26**(11): 7737–7746.
48. Xu, P., et al., Mechanistic understanding of surface plasmon assisted catalysis on a single particle: Cyclic redox of 4-aminothiophenol. *Scientific Reports*, 2013. **3**(1): 2997.
49. Sun, M., et al., Plasmonic scissors for molecular design. *Chemistry: A European Journal*, 2013. **19**(44): 14958–14962.
50. Dong, B., et al., Substrate-, wavelength-, and time-dependent plasmon-assisted surface catalysis reaction of 4-nitrobenzenethiol dimerizing to p,p′-dimercaptoazobenzene on Au, Ag, and Cu films. *Langmuir*, 2011. **27**(17): 10677–10682.
51. Dong, B., et al., Is 4-nitrobenzenethiol converted to p,p′-dimercaptoazobenzene or 4-aminothiophenol by surface photochemistry reaction? *Journal of Raman Spectroscopy*, 2011. **42**(6): 1205–1206.
52. van Schrojenstein Lantman, E. M., et al., Catalytic processes monitored at the nanoscale with tip-enhanced Raman spectroscopy. *Nature Nanotechnology*, 2012. **7**(9): 583–586.
53. Sun, M., et al., In-situ plasmon-driven chemical reactions revealed by high vacuum tip-enhanced Raman spectroscopy. *Scientific Reports*, 2012. **2**(1): 647.

54. Xie, W. and S. Schlücker, Hot electron-induced reduction of small molecules on photorecycling metal surfaces. *Nature Communications*, 2015. **6**(1): 7570.
55. Kim, Y., D. Dumett Torres, and P. K. Jain, Activation energies of plasmonic catalysts. *Nano Letters*, 2016. **16**(5): 3399–3407.
56. Tijunelyte, I., et al., Nanoplasmonics tuned "click chemistry." *Nanoscale*, 2016. **8**(13): 7105–7112.
57. Kazuma, E., et al., Real-space and real-time observation of a plasmon-induced chemical reaction of a single molecule. *Science*, 2018. **360**(6388): 521–526.
58. Prodan, E., et al., A hybridization model for the plasmon response of complex nanostructures. *Science*, 2003. **302**(5644): 419–422.
59. McMahon, J. M., et al., Modeling the effect of small gaps in surface-enhanced Raman spectroscopy. *The Journal of Physical Chemistry C*, 2012. **116**(2): 1627–1637.
60. Hao, F., et al., Plasmon resonances of a gold nanostar. *Nano Letters*, 2007. **7**(3): 729–732.
61. Nordlander, P., et al., Plasmon hybridization in nanoparticle dimers. *Nano Letters*, 2004. **4**(5): 899–903.

Index